AF306998

Wolfgang Reisig

A Primer in Petri Net Design

With 139 Illustrations

Springer-Verlag

Berlin Heidelberg New York
London Paris Tokyo
Hong Kong Barcelona
Budapest

Wolfgang Reisig
Technische Universität München, Institut für Informatik
Postfach 20 24 20, W-8000 München, FRG

Editors

Steven S. Muchnick
SUN Microsystems, Inc., Mountain View, CA 94043, USA

Peter Schnupp
InterFace Lavarde GmbH, Riedgaustr. 13, W-8000 München 80, FRG

This book is based on a German edition published in the series Springer Compass:
W. Reisig: *Systementwurf mit Netzen*
ISBN-13:978-3-642-75331-2

Cover picture: Pastel on paper , John Pearson

ISBN-13:978-3-642-75331-2   e-ISBN-13:978-3-642-75329-9
DOI: 10.1007/978-3-642-75329-9

Library of Congress Cataloging-in-Publication Data
Reisig, Wolfgang, 1950– [Systementwurf mit Netzen. English] A primer in Petri net design/Wolfgang Reisig.
p. cm. – (Springer compass international) Includes bibliographical references and index.
ISBN-13:978-3-642-75331-2   1. System design. 2. Petri nets.
I. Title. II. Series. QA76.9.S88R45 1992 005.1'2'028–dc20 91-37963

Typesetting: Camera-ready by author

45/3140 5 4 3 2 1 0 – Printed on acid-free paper

Preface

This book is about system design using Petri nets. Here, systems are understood very generally as organizational systems in which regulated flows of objects and information are significant. Petri nets have proven to be very reliable in practice for the design of such systems. In the course of the book, we will encounter various net models appropriate for modelling specific systems. They are interrelated in the sense that they have common interpretation patterns and, as a whole, can be understood as a method for representing any given system to any given degree of refinement. The simple and immediately understandable principles of system modelling with nets make it possible to provide an illustrated description of this method without having to go into the mathematics behind it.

The scope of applications of Petri nets has increased since a first edition (in German) of this text appeared in 1985. A whole range of software tools are now available and a lot of case studies can now be referred to. However, aside from these additions, the core text representing Petri nets as a method for system modelling remained valid and is essentially unchanged. The original text was based on courses the author developed for project engineers and project managers in the area of embedded computer systems. I would like to thank the participants in these courses, G. Feistel and H. Keil from the Siemens School for Microcomputers in Düsseldorf, and W. Brauer and P. Schnupp for numerous suggestions, and Franz Goltz for the original illustrations.

I thank the Gesellschaft für Mathematik und Datenverarbeitung (GMD), the German national research institute for computer science, for their support in translating and preparing this new book. I particularly thank Lary Fischer for his careful effort in translating the text, Ms. Elisabeth Münch for redrawing the figures, Ms. Gertrud Jacobs for preparing the layout, and Springer-Verlag for their expert help in the production of the book.

Munich, October 1991   Wolfgang Reisig

Editor's Foreword

In the 1960s there was at least one large blackboard in every room in which software developers worked. It was generally covered with a wide variety of sketches: informal box diagrams, flow charts, transition nets, state diagrams and many others. In those days of batch computing when programmers had to wait for their program listings, they spent a considerable amount of time at the blackboard discussing their programs on the basis of diagrams, improving them and, finally, writing them down so that they were halfway neat and could be used as a basis for discussions with users, for the illustration of problem analyses or specifications and, finally, as an important part of program documentation. This had many advantages:

- People understood each other more rapidly and better in discussions, since they always had a picture of what the other meant and what he was talking about.

- Meetings with users were more pleasant and more effective, since the diagrams at least enabled them to imagine what the systems analysts were getting at with their flowcharts. Difficulties in communication still existed between users and system designers. However, they were less problematic and less time-consuming.

- Specifications and documentations were easier to read, less tiring and, most important, more easily understandable. What one had failed to understand on the basis of a verbal or formal description was perhaps understandable on the basis of the illustration. I am certainly not the only one who, in looking at a chart, noticed that I had totally misinterpreted the author's text and his intentions.

Things have changed since then. The disappearance of blackboards has led to the disappearance of diagrams in most presentations, specifications and software descriptions. This was something that could only be of interest to extremely formal thinkers. Individuals of a more practical bent do not see this as progress in software technology.

What is behind this change (for the worse presumably)? There are several reasons:

- The batch computing situation developed into on-line programming. This resulted in an elimination of waiting times and. in most cases, blackboards. They were replaced by terminals. The latter were "unable to draw".

- In the early 1970s "structured programming" and "stepwise refinement"
  methods gave rise to a new software technology. It unveiled the old flowcharts
  as perhaps not the root of all evil but of quite a few evils. Rightfully so, since
  they were, after all, unstructured. Unfortunately, many computer scientists also
  tended to reject anything that looked faintly like a flowchart. Not rightfully so.
  When, in the early 1980s, university professors maintained that Petri nets were
  the same as flowcharts, we can hardly blame programmers for not knowing what
  to think.
- Most theoreticians didn't like chart techniques anyway. They said they had
  neither syntax nor semantics. What they meant by this was that it was impos-
  sible to prove their formal correctness or to establish precise mapping rules with
  which, at least in principle, one could automatically "transform" a specification
  into an algorithm or verify a program section.

Petri nets, the subject of this book, or somewhat more generally, channel-agency
nets, were already around in the early 1970s. These nets do not have the above-
mentioned disadvantages.

- Software systems have now been developed that permit the user to represent
  Petri nets graphically, store them, edit them, check them and revise them.
- Channel-agency nets are "structured". From the outset, they support stepwise
  refinement better and more systematically than many design and programming
  languages do.
- Their graphic syntax is precisely defined (they are strict "bipartite graphs").
  Semantics exist for them: for the practical programmer the implementation-
  related model of the finite automaton and for the theoretician as much underlying mathematics as he could ever wish for (not the subject of the present
  book).

On the one hand, this mathematical foundation is reassuring. Anyone who uses
Petri nets knows that it is a straightforward method, that it is possible to fall back
on formalisms to clear up doubts and that sufficient theory is available (and in part
implemented on computers) when he needs it.

On the other hand, the considerable amount of theory involved is a disadvan-
tage. Almost all the literature that has been published on the subject thus far is
too mathematical and not practice-oriented enough for the programmer who needs
graphic models and techniques primarily as an illustration for himself, his colleagues
and the users of his software. As such, for more than ten years these nets were almost a secret lore known only to a few systems analysts and software engineers who
had accidentally come across them and had gladly included them in their repertoire
because of their illustrative potential and their precision.

As one of these persons I can say from my own experience that we were not
happy with the fact that "our" analysis and specification technique was so little
known. An important function of any software engineering method is the support
of coordination and information exchange between users, programmers and those

persons or organizations who request program development. For this to be the case, the method has to be known.

As such, I am particularly happy that this series now includes a practice-oriented book on nets. It is a book that can be recommended to the interested reader. More than this, it is a book I would like anyone to have read with whom I will have to specify and plan software in the future. It is a book that can make a very practical design method as well known as it deserves to be. Finally, it is a book that can help solve one of the central problems of application-oriented software development – the improvement of communication between the parties involved.

Peter Schnupp

Contents

Introduction

Why Petri Nets?

Petri nets have been and are being used in many areas of data processing for the modelling of hardware, communication protocols, parallel programs and distributed data bases, particularly in the requirements engineering context, i.e. in the initial phase of system design. The book focuses on this area of application.

What Systems Are Modelled with Petri Nets?

In this book the term "system" does not refer exclusively to computers. Instead, it implies organizational, i.e. logistic, technical and computer-integrated systems of all kinds in which regulated flows of objects and information are of significance. When a computer system is installed in a specific context, there is a need to be able to model general systems of this kind. Every computer is integrated in the environment in which it is intended to serve. However, the service it is to provide can only be described by taking into account a number of components in the environment surrounding the computer.

What Rôle Do Petri Nets Play in System Design?

When a system is being planned or an existing system analyzed, e.g. as a result of the fact that it is to be reorganized, it frequently is the case that

- the system is incompletely and unclearly described (usually in everyday language) and, in addition,
- a software customer is involved in planning an analysis but is unable to handle formal representations.

This can be regarded as the standard situation in system design. As a rule, the procedure in system design is, first of all, to break the system down into rough components, the functions and interrelationships of which can be described in detail. After that these individual components are broken down further to include more detail. Dynamic behavior is defined more and more in the design phase as well. The overall design process must be systematically documented. It has proven to be very advantageous to carry out plausibility and correctness checks as early as possible, since the correction of errors stemming from the design phase is very costly if not done immediately.

The key questions asked in this book derive from the "logic" of system design and its prerequisites:

- How is it possible (in a team or with the customer) to clarify system design questions in a non-formal and yet precise manner?
- How can an informally or incompletely described real system be broken down in a meaningful manner and its components individually analyzed?
- How can we move from informal system representations to descriptions that can be used as a basis for formalization?
- How do we handle systems that are hierarchically structured or are made up of several components that are relatively independent of one another?
- How can we separate different views and relate them to one another at different levels of abstraction?
- How can we model the (dynamic) behavior of a system alongside its (static) structure at the highest possible level?
- How do we represent individual and interrelated processes in (distributed) systems?

Petri nets attempt to give a satisfactory answer to these questions.

What Are the Alternatives?

The field of software engineering emerged in response to the much-discussed "software crisis". Using life-cycle models, this system proposes procedures that systematize the route from the formulation of the problem to a viable program. The initial phases of such procedures are devoted to the assessment and formulation of system requirements. This is referred to as "requirements engineering". A number of methods already exist for requirements engineering, most of them integrated in tools. Petri nets have some similarity with those methods. However, there are major differences.

A detailed comparison of those methods with Petri nets goes beyond the scope of this book. The most obvious difference between Petri nets and other methods is the fact that nets give the same treatment to active and passive components and, above all, they provide the possibility of progressing, at a high level and with any degree of precision, from the description of static components to the representation of dynamic behavior.

The Annual International Conferences on Applications and Theory of Petri Nets provide presentations of such tools. References to such systems can also be found in the periodically published *Petri Net Newsletter* (see bibliography).

What Procedure Does this Book Follow?

This book discusses the concepts and techniques essential to the design of computer-aided systems using Petri nets. The intention is to provide the reader with the basic knowledge he needs to understand net representations, interrelate them and successfully design them himself. For this purpose it provides a systematic presentation of essential design methods and illustrates them with of numerous examples. No mathematics is needed. This is because Petri nets were specially developed with an eye to intuitive understanding and, indeed, are intended to promote and formulate intuitive knowledge. This is necessary, since an essential part of system design comprises

communication between the customer and the design team. The customer cannot
and should not be expected to deal with a complicated formal system. Mathematics
is indispensable where analysis procedures are used to establish the characteristics
of a design. However, procedures of this kind will not be dealt with here.

The book begins with a chapter of examples that illustrate the essential principles
of system design with nets. After that it moves systematically from special to higher
net models and, finally, arrives at a set of guidelines for system design with nets.

To be sure, practice is needed to apply the concepts presented. Various smaller
examples, exercises, and, in particular, an "extensive" example of a wholesale company in the last chapter serve this purpose.

What Objective Does the Book Pursue?

The objective of this book might be illustrated on the basis of a comparison with
getting a driver's license. The driving school (this book) conveys basic knowledge
and provides guidelines for appropriate use of a vehicle (system design with nets).
The beginner takes driving lessons (exercise classes) to practice using the vehicle
under supervision and in realistic situations. At the end of the driving training
period the beginner becomes a driver (Petri net user) who then develops his own
routines, preferences and style. He soon knows how to deal with new problems that
crop up. Just as a driver does not need to know much about the technical side of
his car, a Petri net user does not have to have a profound knowledge of net theory.
This is only necessary if someone wants to check for correctness system designs
represented as nets.

Sticking with the same comparison, this book may be thought of as a kind of
driver's manual for Petri net users.

1 Principles of System Design with Nets

1.1 An Example

Taking the organization of a library as an example, let us begin with the most basic meaningful structure of a library. The users of the library have access to a stock of books kept in the *stacks*. Thus, we are dealing here with two components. It would seem obvious enough to describe one of the components (the stock of books) as being *passive*, and the other (the users) as being *active*. If we represent the passive component with a *circle* and the active component with a *box* we have the configuration shown in Fig. 1. The arrows between the components indicate the flow of objects and information.

In large libraries a user may not have free access to the stacks. Instead there may be one (or several) counters at which users are served by library staff. Intuitively we understand which components are passive (the counters) and which are active (library staff). Based on this information, we can move on to the organizational structure indicated in Fig. 2.

Fig. 1   Simplest meaningful description of a library

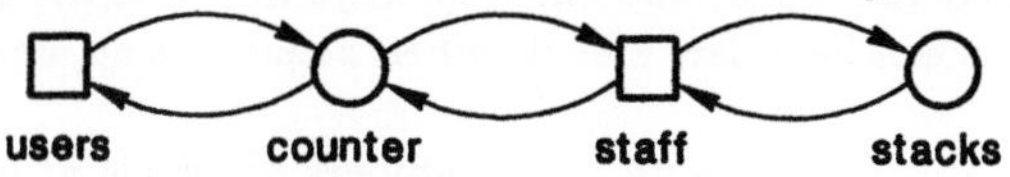

Fig. 2   Organization of a large library

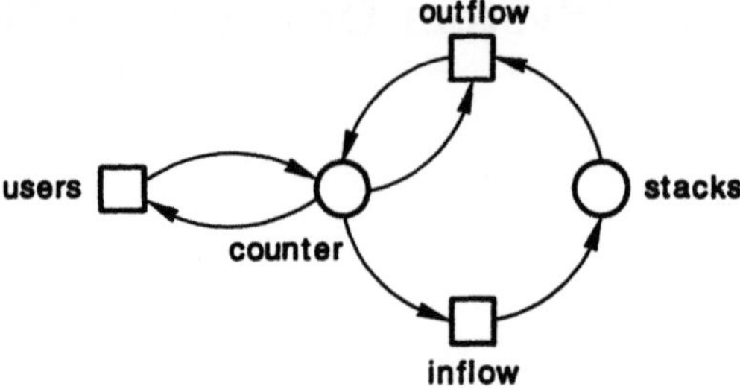

Fig. 3 Organizational distinction between outflow (lending) and inflow (return) of books

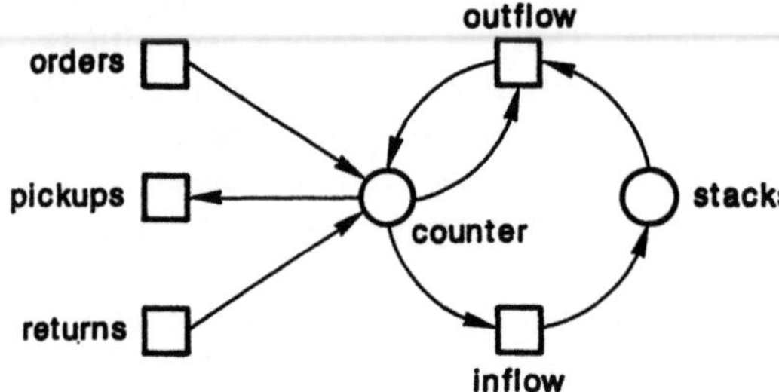

Fig. 4 Breakdown of user interactions with the library into the acts of ordering, picking up and returning books

A large volume of book borrowing requires organization. We first of all need to distinguish the lending (outflow) and return (inflow) of books. In Fig. 3 we assume that before books can be borrowed they first of all have to be ordered. In other words, there has to be a flow of information from the borrower to the lender before a given book will be transferred from the lender to the borrower. In this connection, it would seem useful to identify the different ways in which the borrower interacts with a library, i.e. when he orders a book, when he picks it up, and when he returns it (see Fig. 4).

In order to be able to deal with large numbers of people at one time, it would seem sensible to establish separate counters for ordering, picking up and returning books (Fig. 5).

The library keeps a record of the books borrowed, in the simplest case by using index cards that are stored with the books in the stacks. Whenever a book is borrowed, its card is placed in a borrowed-book file. When the book is returned, the relevant card is taken out of the file and returned to the stacks along with the book it belongs to. Fig. 6 adds a borrowed-book file to the system. (This would only be the first step if we wanted to plan the computer-aided reorganization of a manually administered library.)

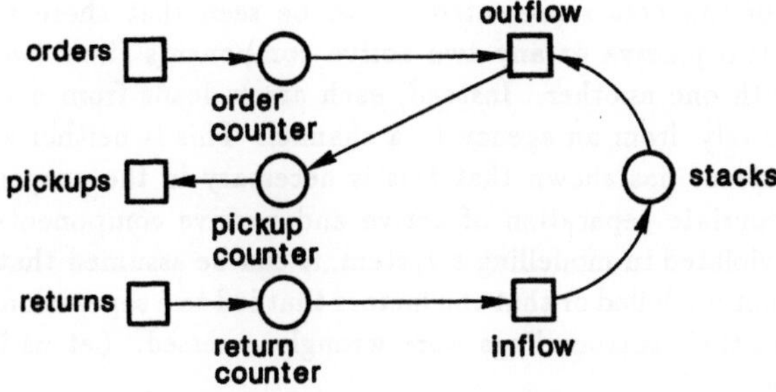

Fig. 5   Introduction of separate counters

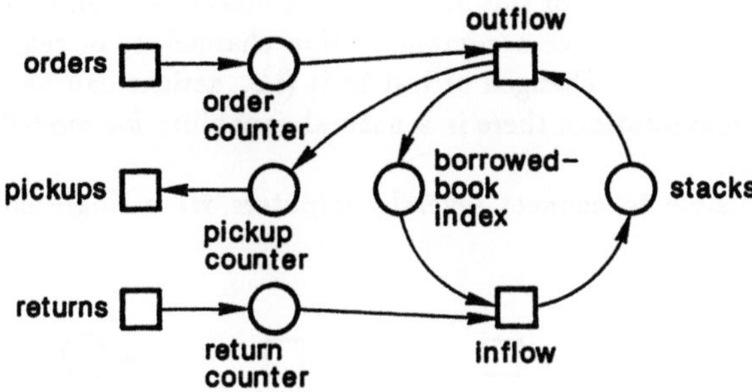

Fig. 6   Introduction of a borrowed-book index

1.2   Passive and Active Components

The library example has been developed far enough for our purposes. We can now use it to illustrate a number of the central principles of system design with nets. The first thing that needs to be done is to distinguish components of the system and to designate every component as being either passive or active. In our example this was quite obvious. The passive components (i.e. the various counters, the stacks and the borrowed-book index) are able to store things or make them visible. They can be in different states. They are referred to as *channels*. The active components (in our example the library users in their three interactive roles, i.e. ordering, picking up and returning books, as well as the act of retrieving books from the stacks and the act of putting them back on the shelves) are able to produce, transport or change things. They are referred to as *agencies*. Thus, the above figures represent *nets made up of channels* (circles) *and agencies* (boxes).

It goes without saying that arrows are also important in designing channel/agency nets. An arrow never represents a system component, but rather always an abstract relationship between components, e.g. logical connections, access rights, physical proximity or direct links.

In the case of the nets represented it can be seen that there are no arrows connecting any two passive or any two active components (i.e. two channels or two agencies) with one another. Instead, each arrow leads from a channel to an agency or, conversely, from an agency to a channel. This is neither accidental nor arbitrary. Experience has shown that this is necessary in the proper use of nets, i.e. in the appropriate separation of active and passive components. Whenever this principle is violated in modelling a system, it can be assumed that either a real component was not modelled or that the factors that led to a separation of individual components from their surroundings were wrongly assessed. Let us illustrate this with an example.

Fig. 7 shows a wrong and a right way (i.e. in terms of Petri nets) of modelling the communications channel in a computer link. In the drawing on the left the communications channel is either not modelled as a separate component, or it has been assumed that channels of this kind always connect two computers. Both may prove to be impractical, since a communication channel is, in reality, a complex component. Data may be changed or lost in it (i.e. actions can have an influence on it). In net representations there is a natural capability for modelling actions of this kind (Fig. 8).

It is also possible to connect several computers via a single communications channel (Fig. 9).

Fig. 7   Representations of a simple computer link

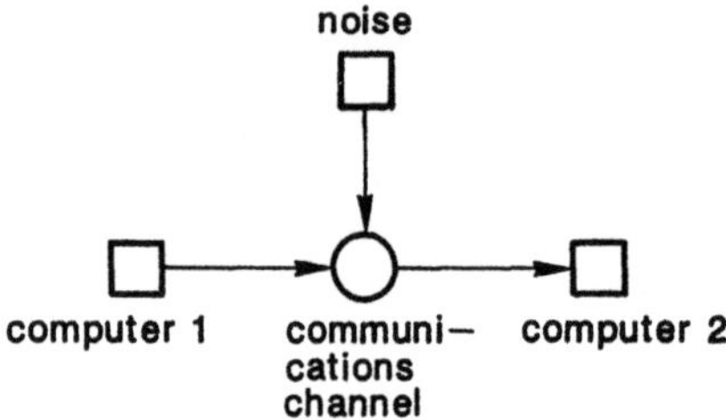

Fig. 8   Extension of Fig. 7

Fig. 9   Several computers linked via one channel

1.3   The Change to Dynamic Behavior

At this point we should focus on another central principle of system design with nets, i.e. that of systematically moving from nets made up of channels and agencies, such as the one we have just dealt with, to nets that model dynamic behavior. Let us return to our library example for this purpose, making use of the structure we established in Fig. 6. Fig. 10 shows that the channels contain concrete objects, i.e. an order form (filled out) is at the order counter, a book is at the pickup counter waiting for the person who ordered it, and at the return counter there is a book that has been returned but not yet put back on the shelf. The books in the stacks and the cards in the borrowed-book index are also indicated. The agencies can now redistribute the objects, acting in accordance with rules formulated by the system designer. Order forms are accepted by the lending agency, books are retrieved from the stacks and brought to the pickup counter. Borrowers pick up the books they have ordered from this counter.

Returned books are put back in the stacks. How this dynamic behavior is represented is explained in Chaps. 2-4.

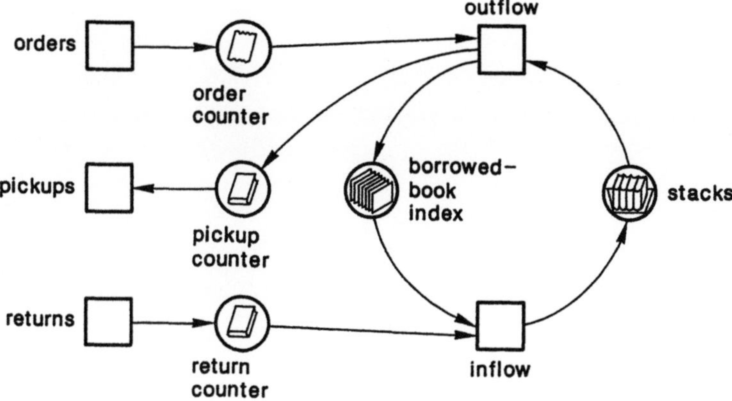

Fig. 10   Order forms, books, index cards

1.4 Relationships Between Net Representations

The last principle we want to deal with here concerns the relationship between different net representations. By way of example, the change from Fig. 2 to Fig. 3 involves a more detailed description of the functions exercised by library staff. A similar relationship exists in the changes from Fig. 3 to Fig. 4 and from Fig. 4 to Fig. 5. In these cases more detailed descriptions were given of the interactive roles played by library users and by the library itself (counters). The change from Fig. 5 to Fig. 6, on the other hand, is different. Here, a new component is added to the model (the borrowed-book index). Thus, there are two ways of developing a model, i.e. either by *replacing* a component with a more detailed subnet or by *adding* components to the system. These techniques are dealt with in Chap. 5.

The three principles that have been presented here, i.e. breaking the system down into passive and active components, the transition from the static nature of separate components to the dynamic behavior of a system as well as the interrelating of individual net representations, form the basis of an integrated technique for designing systems with nets.

2　Condition-Event Nets

2.1　An Example

We will begin with a system that could come from a wide range of different areas. In it objects are generated, stored in a channel, removed at a later point in time and, finally, consumed. In a specific instance, 'storing in a channel' could also stand for 'dispatch', 'make available' or 'hand over'. 'Remove from the channel' could also mean 'accept' or 'receive'. The objects could be goods, news items, data messages, money or even services. We are not interested here in a specific possibility, but rather in what they all have in common.

In Fig. 11 'dispatch' and 'retrieval' are *events* in our (planned) system that may be *recurrent*. The event 'dispatch', for instance, may occur if certain preconditions have been met, i.e. if the producer is ready to dispatch and the channel is empty. When the event 'dispatch' takes place the producer is then ready to produce and the channel is occupied. 'Retrieval' is an event dependent on two conditions, i.e. the readiness of the consumer to retrieve and the occupied state of the transmission channel.

The prerequisites for an occurrence of an event are formulated with the help of *conditions*. In a specific situation every condition is either *fulfilled* or *not fulfilled*. The event 'dispatch' may occur if the condition 'producer ready to send' has been fulfilled and the conditions 'channel occupied' and 'producer ready to produce' have not been fulfilled. As a result of the occurrence of 'dispatch' the condition 'producer ready to dispatch' is not fulfilled. The conditions 'channel occupied' and 'producer ready to produce' are fulfilled. The corresponding situation applies in the case of the event 'retrieval'. On its occurrence the conditions 'channel occupied' and 'consumer ready to retrieve', which must first of all have been fulfilled, are now unfulfilled and, conversely, the condition that was initially unfulfilled, i.e. 'consumer ready to consume', is now fulfilled. The following components appear in Fig. 1: conditions ($\bigcirc$), events ($\square$) and arrows. An arrow $b\bigcirc\longrightarrow\square\,e$ indicates that b is a precondition of e, an arrow $e\square\longrightarrow\bigcirc b$ indicates that b is a postcondition of e. The conditions fulfilled in a given instance are designated with a token ($\odot$).

If an event occurs, its previously fulfilled preconditions are unfulfilled and its (previously unfulfilled) postconditions are fulfilled (Fig. 12).

Fig. 11 shows two fulfilled and three unfulfilled conditions and the event 'dispatch' (and only this) can occur. If it occurs, the configuration represented in Fig.

13 is created. Now two further events may occur, i.e. 'produce' and 'retrieve', and this fully independently of one another. As a result of these and of 'consume' the configuration in Fig. 11 is attained once again.

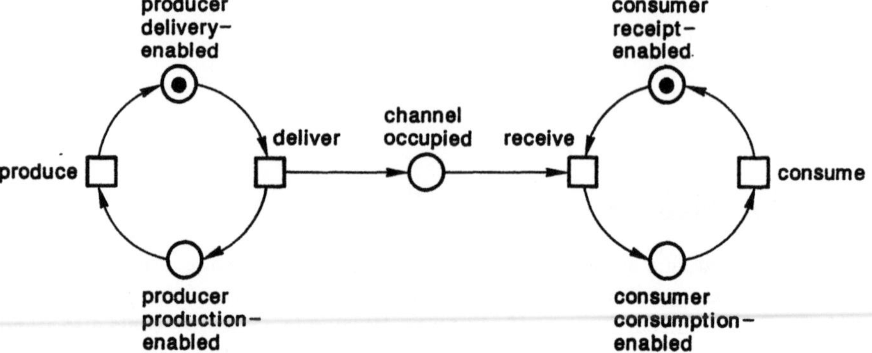

Fig. 11   A producer-consumer system

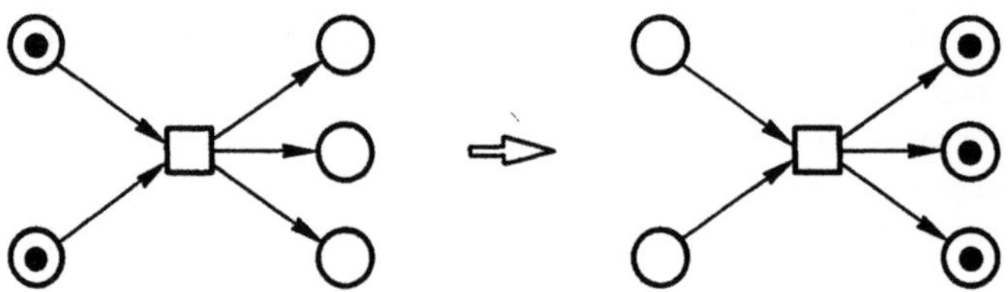

Fig. 12   The effect of the occurrence of an event on its pre- and post-conditions

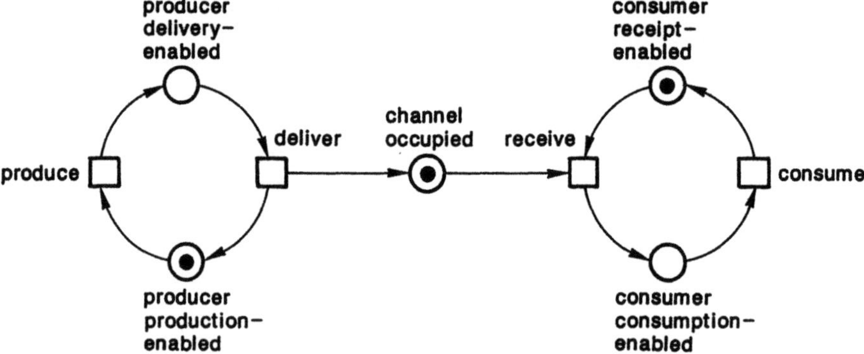

Fig. 13   The configuration deriving from Fig. 11 as a result of the occurrence of the event "produce"

2.2 Rules

We can derive the following rules from the representations in Figs. 11-13:

> A *net consisting of conditions and events* is based on
> - *conditions*, represented as circles ($\bigcirc$);
> - *events*, represented as boxes ($\square$);
> - *arrows from conditions to events* ($\bigcirc\!\longrightarrow\!\square$);
> - *arrows from events to conditions* ($\square\!\longrightarrow\!\bigcirc$);
> - *tokens* in some conditions ($\odot$) which indicate the *initial* case, i.e. the conditions fulfilled at the outset.

> In a net consisting of conditions and events
> - a condition b is a *precondition* of an event e if there is an arrow $b\,\bigcirc\!\longrightarrow\!\square\,e$;
> - a condition b is a postcondition of an event e if there is an arrow $e\,\square\!\longrightarrow\!\bigcirc\,b$;
> - in any given situation every condition is either *fulfilled* or *unfulfilled*;
> - every fulfilled condition is indicated by a *token*;
> - a *case* consists of the conditions fulfilled in a given situation.

Fig. 11 and 13 indicate two different cases of the same condition-event net.

> An event in a condition-event net can occur (in a given case) if all of its preconditions are fulfilled and all of its postconditions are unfulfilled. Such events are (in the given case) said to be *activated.*
> If an event is activated and *occurs*, its preconditions are unfulfilled and its postconditions are fulfilled.

Problem 1
Change Fig. 11 such that the channel will be able to accomodate two objects. (Hint: the channel should now consist of two storage cells in series.)

Problem 2
Fig. 14 shows the system of the four seasons with its cyclical change.

Represent the following conditions:
a) It is autumn or winter,
b) It is not summer.

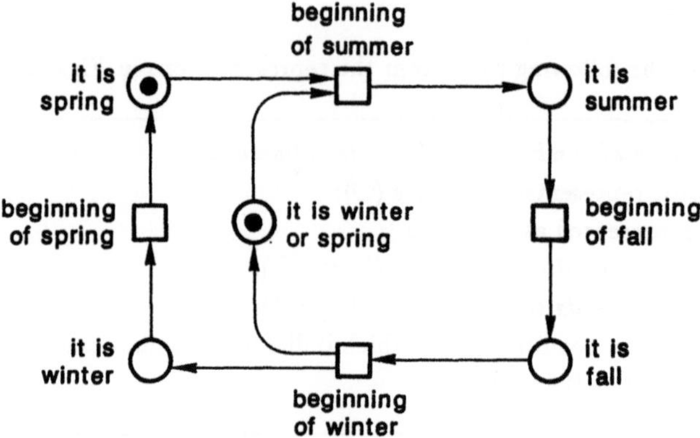

Fig. 14   The four seasons and their changes

2.3   Conflicts

An important characteristic of condition-event nets is their *non- determined behavior
in the case of a conflict.*

As an example consider two (operating system) processes p_1 and p_2, both of
which can access the same storage area.

Let access be arbitrary and without a defined sequence. However, simultaneous
access is ruled out. Fig. 15 shows a net based on this example. Restricted storage
access is brought about with the aid of a 'key' that every process has to carry with
it during storage access and which it returns after storage access. Since there is
only *one* key, the two processes can never have access at the same time. In the case
represented the events 'p_1 takes the key' and 'p_2 takes the key' are both activated.
Both events may occur. However, they cannot occur independently of one another
(such as in the case of 'produce' and 'retrieve' in Fig. 13). If one of the two events
occurs, the other is no longer activated (in contrast to the events mentioned in connection with Fig. 13). The two events *compete* for the available key. They are *in
conflict* with one another.

> Two events of a condition-event net are *in conflict* with one another if both are
> activated and the other is no longer activated as a result of the occurrence of one
> event.

Two activated events are in conflict with one another if they have at least
one pre-condition or one post-condition in common.

Conflicts do not always arise when events have common pre- or post-sets (Fig. 17).

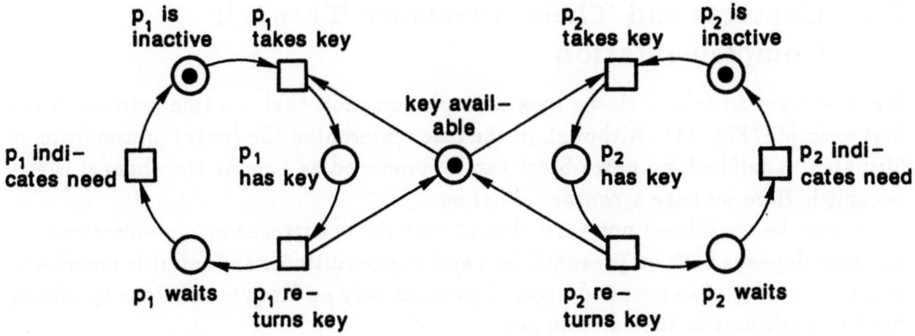

Fig. 15   Two processes with limited buffer access

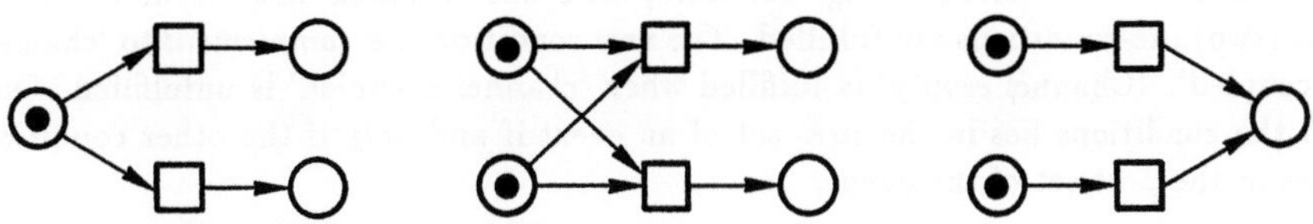

Fig. 16   Examples of conflicts

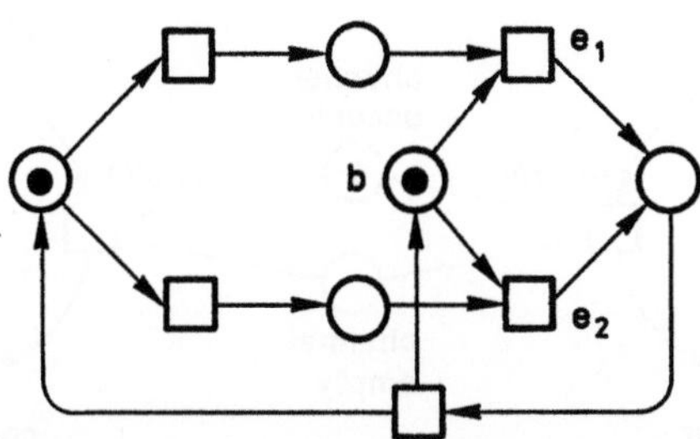

Fig. 17 e_1 and e_2 both have the condition b in their pre-set. However, there is never a conflict between them

Problem 3

a) Represent in Fig. 14 the condition 'it is summer or winter'.
b) Supplement Fig. 15 such that __three__ processes compete for the key.

Problem 4

Change Fig. 11 such that now __two__ channels are available. If both are empty it is not defined which of the two will be occupied on dispatching. If both are occupied, it has not been defined which of the two will be emptied on retrieval. (Hint: Design two different events for occupying the channels and two different events for emptying them.)

2.4 Contacts and Their Avoidance Through Complementation

We now want to take a closer look at a phenomenon that became evident in our first example (Fig. 11). Although in the case represented the (only) precondition of 'dispatch' is fulfilled, no new object can be produced as long as the channel is still occupied. Here we have a *contact situation.*

It may be considered not very elegant that the occurrence or non-occurrence of an event depends both on preconditions and postconditions. Indeed, it is possible to make the ability of an event to occur dependent only on its preconditions by adding further conditions to the existing net.

In our producer/consumer example we attached the occurrence of 'dispatch' (among other things) to the condition that the channel be empty. However, the condition 'channel empty' does not appear in Figs. 11 and 13. In Fig. 18 it is included as an extension of Fig. 13. Here, the event 'dispatch' may occur whenever its (two) pre-conditions are fulfilled. The new condition is a *complement* to 'channel occupied'. 'Channel empty' is fulfilled when 'channel occupied' is unfulfilled. One of the conditions lies in the pre- set of an event if and only if the other condition lies in the post-set of the event.

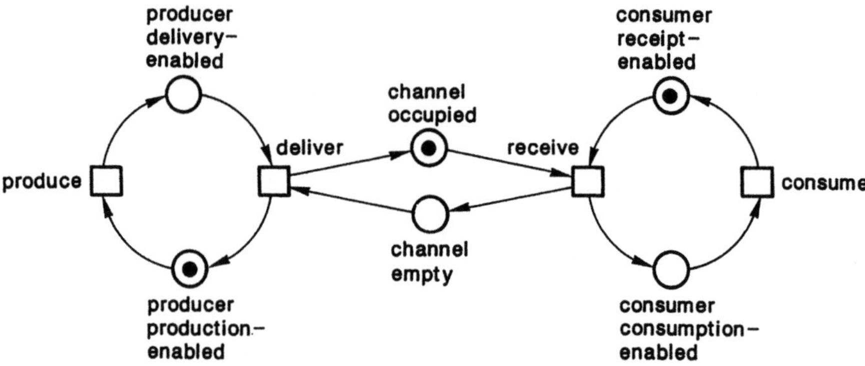

Fig. 18   "Channel empty" as a complement of "channel occupied"

In a condition-event net a condition b' is a *complement* to a condition b if the following is valid for every event e:
- b is a pre-condition of e when b' is a post-condition of e;
- b is a post-condition of e when b' is a pre-condition of e;
- b' is unfulfilled in the initial case if and only if b is fulfilled in the initial case.

The following applies in a condition-event net:
-   If b' is a complement of a condition b, exactly one of the two conditions is fulfilled in each case.
-   If we supplement the net by adding the complement of a condition, this does not change its behavior.

In a condition-event net a *contact* exists if all pre-conditions and at least one post-condition of an event are fulfilled.
A condition-event net is *contact-free* if a contact can never occur.

The net in Fig. 11 is thus not contact-free. Fig. 18 shows a contact-free net.

■ A condition-event net can be made contact-free by constructing complements.

In a condition-event net a complement can be constructed for every condition as long as it is not already present (Fig. 19). If the original net was not contact-free it will become contact-free with the addition of the complements. A net can, however, be contact-free without all of its conditions necessarily being accompanied by a complement. Fig. 17 provides an example of this.

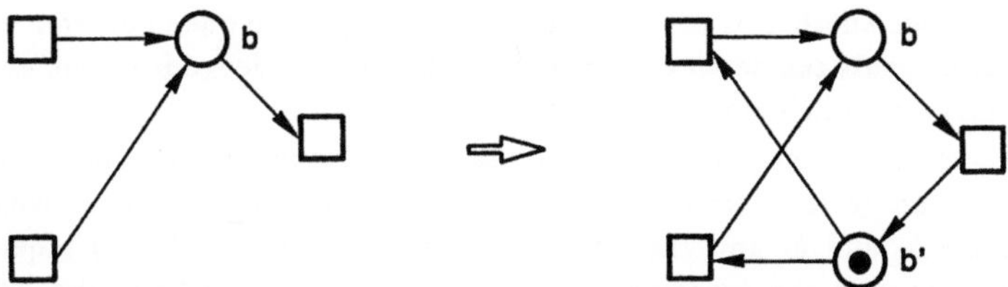

Fig. 19   Construction of the complement b' of a condition b

Problem 5
Are Figs. 14 and 15 contact-free?

Problem 6
Represent the complements of all conditions in Fig. 14. What conditions do they designate?

Problem 7
Represent the complement of the condition 'key is available' in Fig. 15.

2.5 Processes of Condition-Event Nets

In Sect. 2.2 it was stated that every condition-event net includes the indication of an *initial case*. Based on it or another case, events may occur and, as a result, the respective cases may be changed. A complex of events of this kind is referred to as a *process* (we distinguish process in this meaning from the way it was used in Sect. 2.3, i.e. an operating system process is a system, possibly representable as a condition-event net. What is meant in this case is a collection of individual occurrences of events in systems or nets). In a process events can recur and conditions can change repeatedly. One might be tempted to define a process as a sequence of occurring events (the conditions that are changed in this context result from the net). However, a sequence of this kind is not always given. In Fig. 13 a case is represented in which the events 'produce' and 'receive' can occur independently of one another. If both events occur, the representation 'first production, then receipt' is just as right (and wrong) as 'first retrieval, then production'. Fig. 13 does not define either of the two sequences and it is not recommendable to do so in the representation of a process, but rather to note that the two events occurred independently of one another.

One may object that 'in reality' there is always either a sequence or the events occur 'simultaneously'. This view is problematic for two reasons:

1) It presupposes that points in time can be determined and compared with one another at any location and in any situation. Aside from objections that may be made by modern physics (the events in question may occur at locations quite distant from one another or they may move with respect to one another), whether or not two events can be viewed as simultaneous or not depends on the precision of the measuring instruments.

2) If a sequence in the occurrence of events is important for the system, suitable measures can be taken to bring this about. As a result of the complementary conditions b_0 and b_1 and the event e in Fig. 20, an (arbitrary) sequence is brought about in the occurrence of the two events 'production' and 'receipt'. After the occurrence of the first of the two events, e must occur before the other event can occur. It follows from this that wherever sequences are desired or necessary, they can (and should) be imposed. However, where this is not the case, the independent (parallel, concurrent, non-sequential) occurrence of events should be taken note of and described.

How can we represent processes so that the above requirements are fulfilled? We represent every occurrence of an event as a box and every fulfillment of a condition as a circle and indicate (by means of an inscription) what events or conditions are involved. Arrows designate the causal relationship between fulfilled conditions and occurring events. Fig. 21 describes a process based on Fig. 18. Only processes in contact-free nets can be represented in this configuration.

However, in the previous section we saw that any condition-event net can be made contact-free by constructing complements. Thus, the restriction to contact-free nets is not a restriction.

Fig. 22 shows an example of the construction of processes.

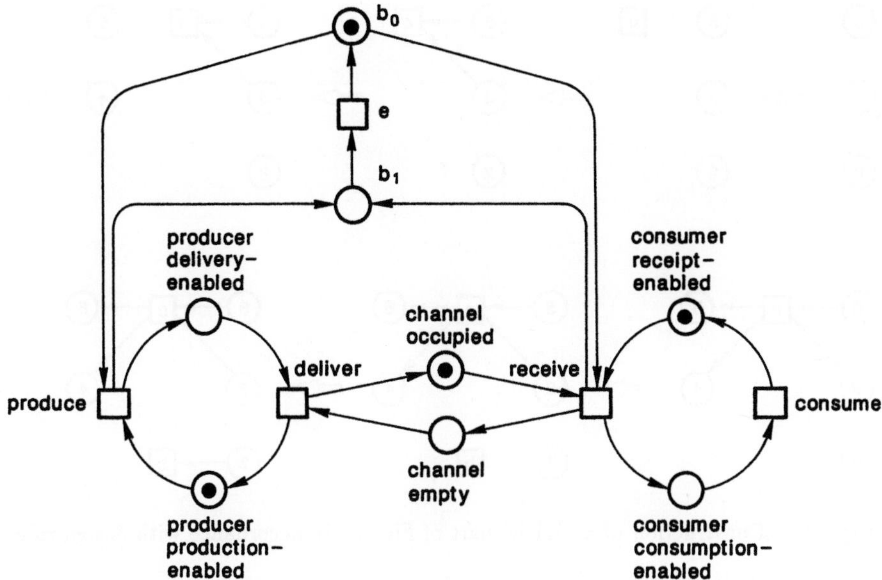

Fig. 20 Addition to Fig. 18 that imposes (arbitrary) sequences in the occurrence of the events "produce" and "receive"

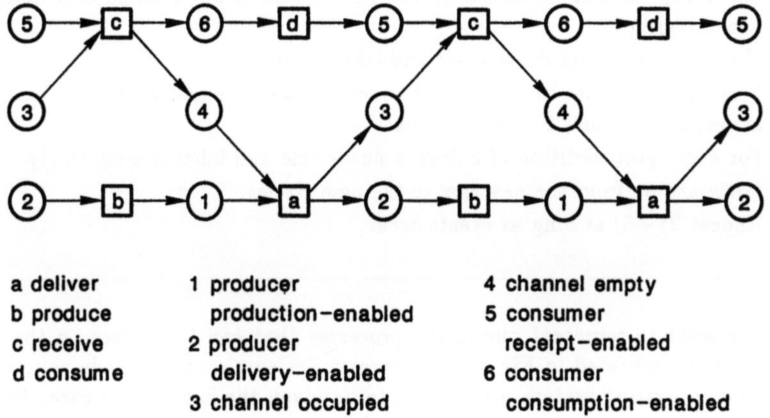

a deliver	1 producer production-enabled	4 channel empty
b produce	2 producer delivery-enabled	5 consumer receipt-enabled
c receive	3 channel occupied	6 consumer consumption-enabled
d consume		

Fig. 21 Process based on Fig. 18

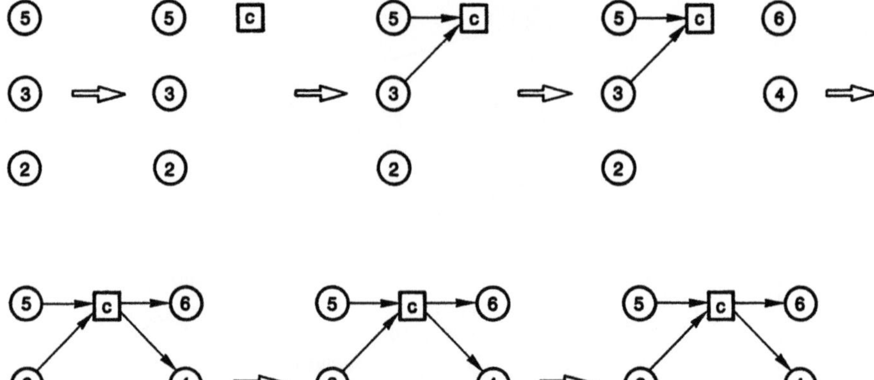

Fig. 22   Construction of an initial part of Fig. 21 in accordance with design rules

The following procedure is used to design a *process in a contact-free condition-event net*:

1) For every fulfilled condition of the initial case draw a circle and label it accordingly.

2) If an event *e* occurs draw a box and label it with *e*.

3) From all the circles inscribed with preconditions of *e* and which still have no arrow, draw arrows to the next box.

4) For every postcondition of *e* draw a new circle and label it accordingly.

5) Draw arrows from the new box to the new circles.

6) Repeat 2) – 5) as long as events occur.

If we want to represent one of the processes that can take place on the basis of the system indicated in Fig. 15, a recurrent decision must be made as to which of the two operating system processes is to be given the key in each case. Fig. 23 shows two of many possible processes based on Fig. 15.

Since every time a condition is fulfilled or an event occurs, a circle or a box is drawn, and since in a process the conflicts of the underlying system are solved in favor of one of the alternatives, it can be said that:

In a process representation:
- there is no arrow sequence that forms a cycle;
- at most one arrow leads to a circle; and
- at most one arrow leads away from a circle.

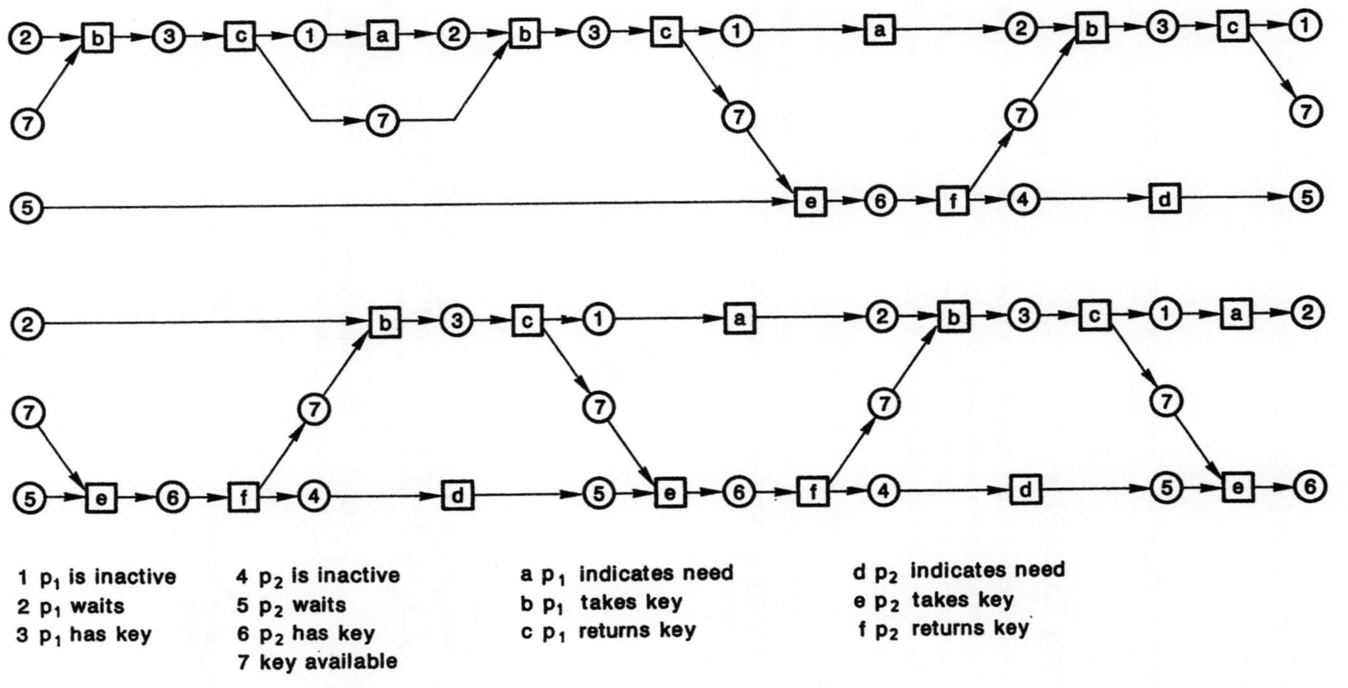

Fig. 23   Two processes of the system given in Fig. 15

Problem 8

Design a process for the net in Fig. 14 and for the supplement as required in Problems 2 and 3a.

Problem 9

Design processes for the net in Fig. 20 and compare them with Fig. 21.

2.6 Further Examples

First of all we will discuss the organization of a small (production) system (Fig. 24). The system consists of three machines M_1, M_2, M_3 and two operators B_1 and B_2. Fill orders on the basis of the following rules: Every order will be processed first by M_1, then by M_2 or M_3. The operator B_1 will work on M_1 and M_2, whereas B_2 will work on M_1 and M_3.

Problem 10

Supplement Fig. 24 by adding the conditions 'M_2 available' and 'M_3 available'. Why isn't it necessary to introduce these conditions? Why, on the other hand, is 'M_1 available' necessary?

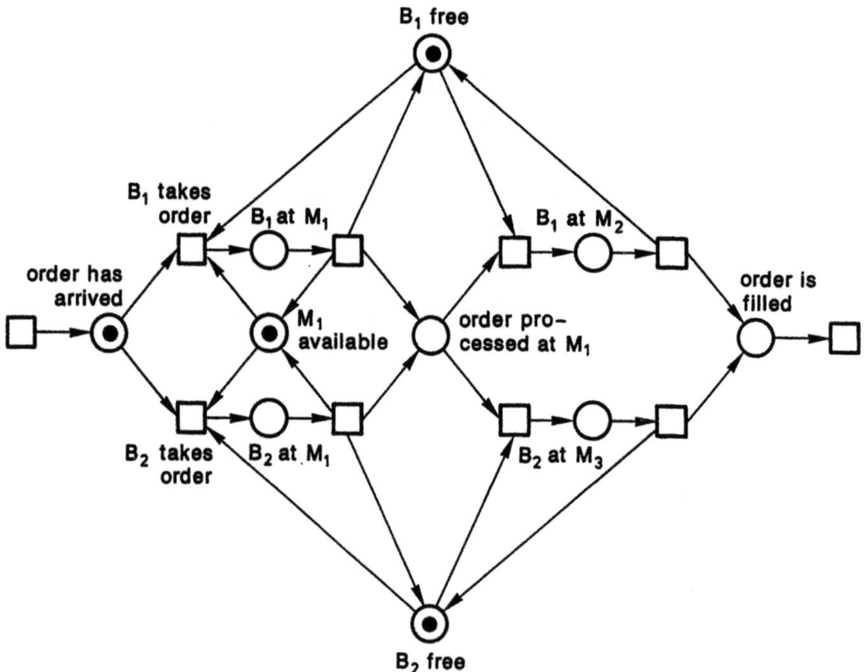

Fig. 24 Organizational diagram of a (production) system

22

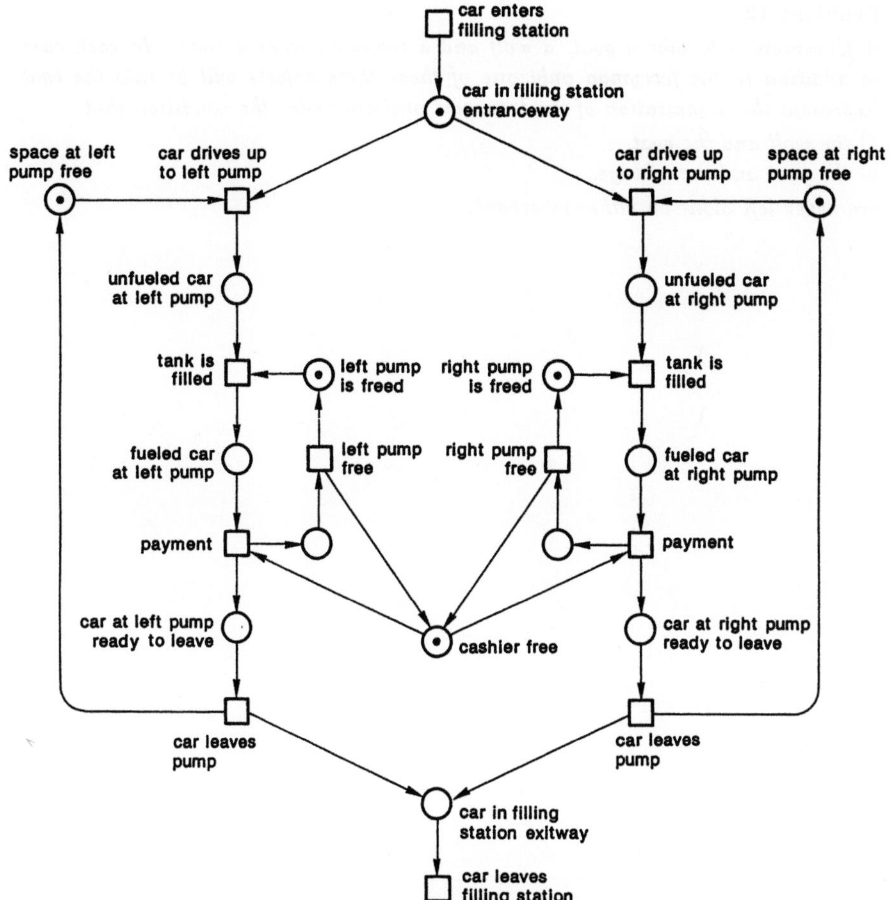

Fig. 25   Organizational diagram of a self-service filling station

We will conclude this chapter with the organizational diagram of a service station. Fig. 25 represents a self-service filling station as a condition-event net with the following characteristics: There are two pumps. There is space next to each pump so that only one car can be fueled when parked in this space and when the pump is available (green light). There is an attendant who is paid by the customer before the respective pump is cleared for use by the next customer.

Problem 11

Change Fig. 25 such that

a) *There are two spaces next to each pump;*

b) *Two attendants are employed who can take money at both pumps and clear them both.*

Problem 12

A ferryman is to take a goat, a wolf and a cabbage across a river. In each case, in addition to the ferryman only one of these three objects will fit into the boat. Represent the organization of the transport problem under the condition that

1) the wolf and the goat,
2) the goat and the cabbage,

are never left alone on either riverbank.

3 Place-Transition Nets

3.1 An Example

In Problem 1, Fig. 11 was to be changed such that two objects can be accomodated in the channel. This can be done without any great difficulty. However, this design would be very cumbersome for, say, 10 or 30 objects (Fig. 26).

In our place-transition net representation every dispatched object is still indicated in storage as a token but now the channel itself is represented as a single circle. We extend the rules to admit up to 10 tokens in this circle. Every event 'dispatch' increases the number of tokens by one and every event 'removal' reduces this number by one (Fig. 27). In contrast to Fig. 26, the precise location of the objects in channel storage cells is no longer visible in Fig. 27.

In Fig. 28, a second consumer is added to the system in Fig. 27. As with the objects produced in Fig. 27, it may be the case that we only want to take the number of consumers into consideration without distinguishing them individually. In this case the two consumers can be placed together in a single net component (Fig. 29).

The same method can be used to represent the fact that two producers and three consumers are involved in the system (Fig. 30).

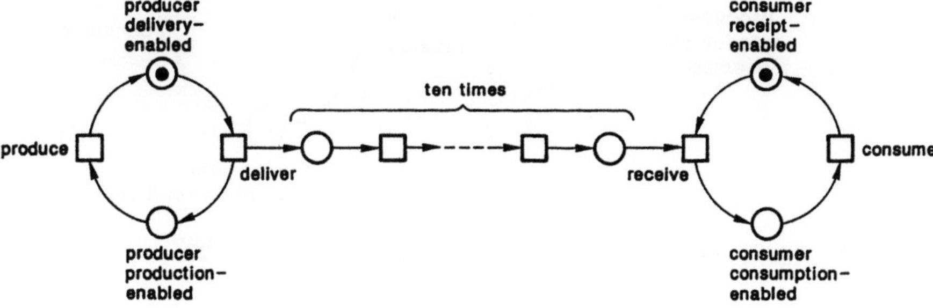

Fig. 26   Change of Fig. 11. The channel can now accomodate up to 10 objects

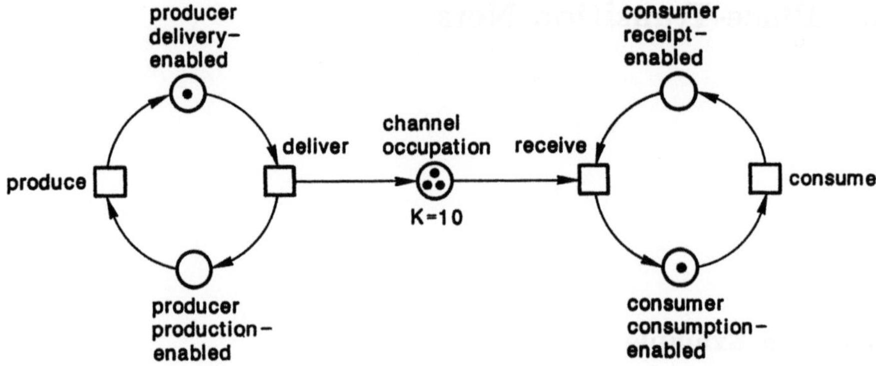

Fig. 27 Concentrated representation of Fig. 26 with three objects in the channel

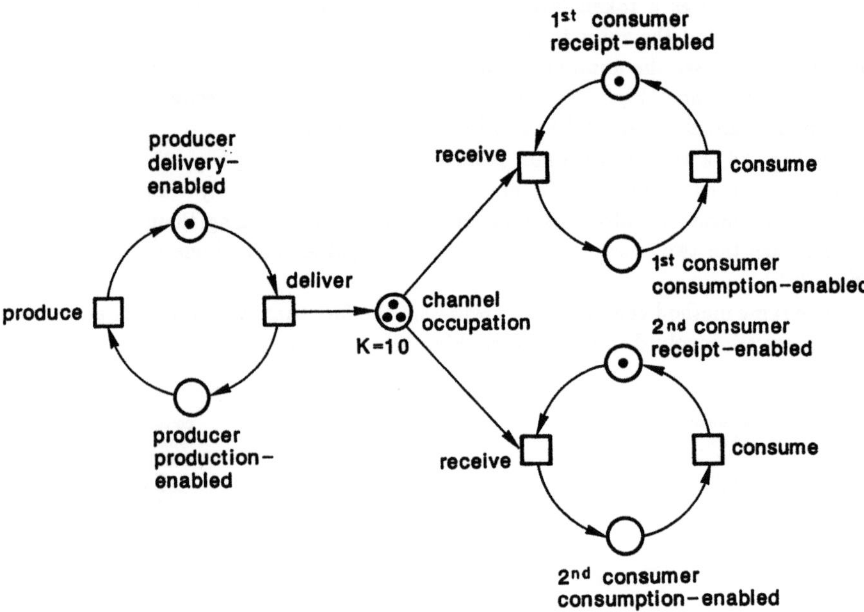

Fig. 28 Addition of a second consumer to the system in Fig. 27

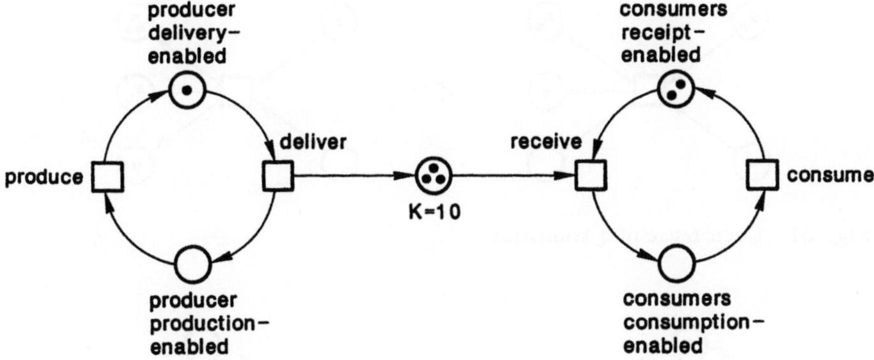

Fig. 29   Concentration of the consumer component in Fig. 28

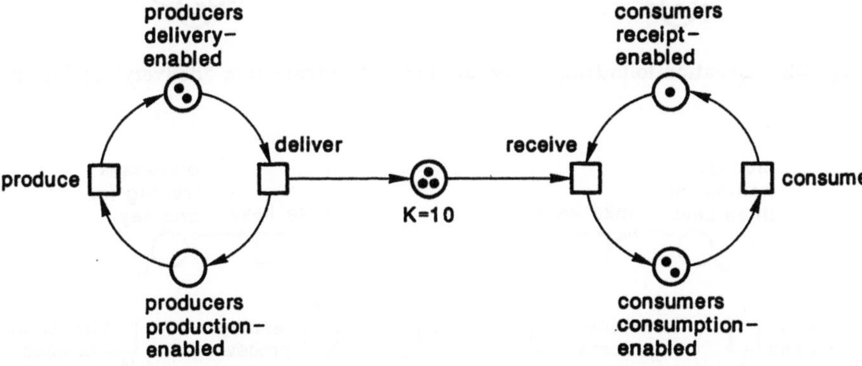

Fig. 30   Modification of Fig. 29, now involving two producers and three consumers

Now, we can no longer speak of 'conditions' and 'events'. Instead, we refer in a more general (i.e. a more abstract) sense to the circles as *places* and the boxes as *transitions*. We explain dynamic changes on the basis of the *occurrence* of transitions in accordance with the rule indicated in the example in Fig. 31.

Based on this rule, the situation in Fig. 32 is brought about by the occurrence of 'deliver' in Fig. 30.

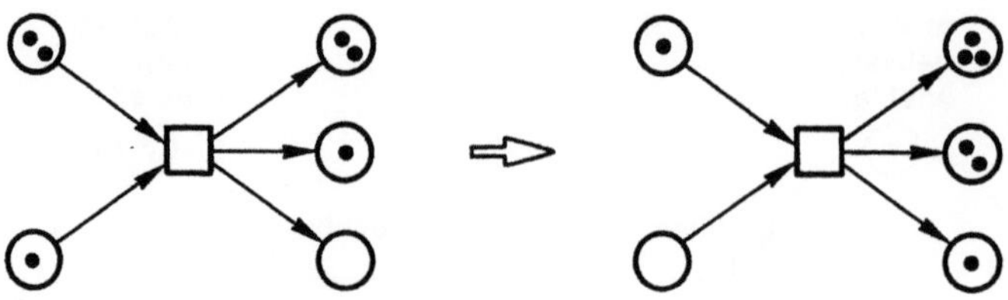

Fig. 31   Occurrence of a transition

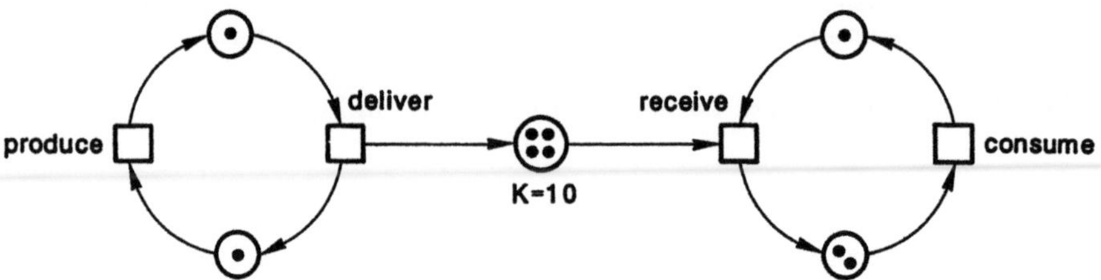

Fig. 32   Situation following the occurrence of the transition "delivery" in Fig. 30

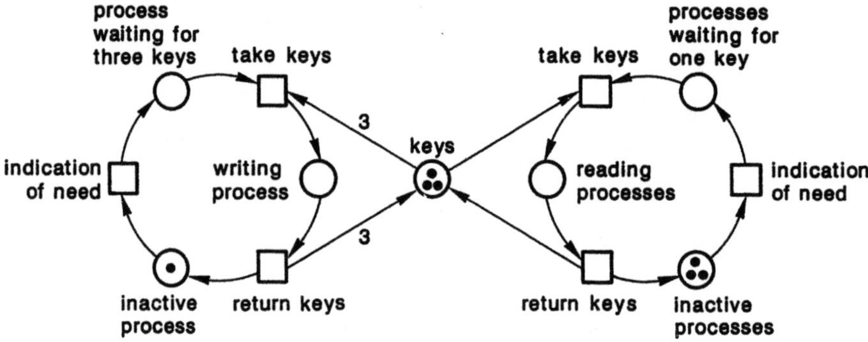

Fig. 33   A system consisting of one write- and three read- authorized processes

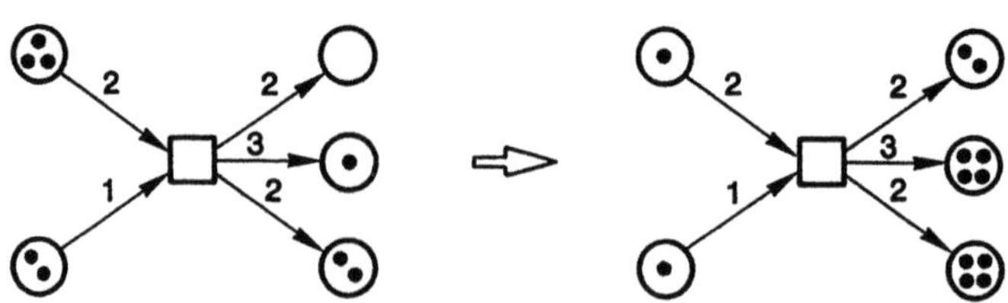

Fig. 34   Occurrence of a transition with arrow weights

3.2 Arrow Weights

As a further example let us look at a modification of Fig. 15. We will now assume
three read-storage processes and a fourth that modifies storage contents. Obviously
it is sensible to give read-authorized processes independent (parallel, concurrent)
access to storage. The write-authorize process should, of course, only be able to
access storage when none of the storage-read processes are active.

Fig. 33 represents this system. Three keys are used. To read we need one key.
However, to write we need all three. This is expressed by means of an *arrow weight*
'3'. A number label on an arrow indicates that on occurrence of the respective transition as many tokens 'flow through the arrow' as indicated by this number. The
example in Fig. 34 shows how a transition occurs when arrow weights are involved.

3.3 Rules

A *place-transition net* consists of
- *places*, represented as circles (O);
- *transitions*, represented as boxes (□);
- *arrows from places to transitions* O——▢ ;
- *arrows from transitions to places* □——O ;
- a *capacity indication* for every place (represented as label K =...);
- a *weight* for every arrow (represented as a number);
- an *initial marking*, defining the initial number of tokens for every place
  (cannot be greater than the indicated capacity).

The arrow weight '1' can be omitted. The capacity of a place does not need to be
indicated if it is not important or if there is never a danger that it will be exceeded.
A place can have an unrestricted (infinite) capacity.

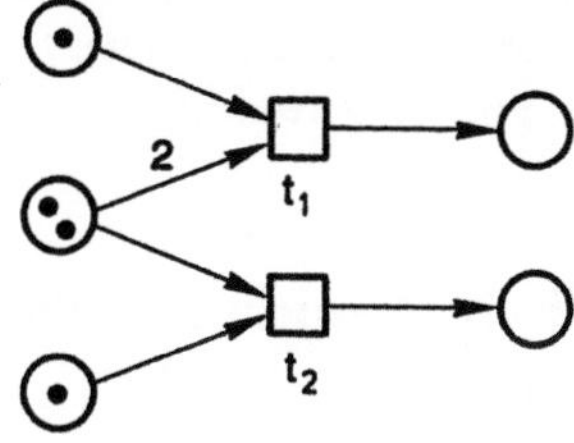

Fig. 35   A conflict between t_1 and t_2

In a place-transition net
- a *marking* is indicated by the number of tokens in every place;
- a place p is in the *pre-set* (or *post-set*) of a transition t if there is an arrow
  ($p\bigcirc\!\!\longrightarrow\!\Box\, t$ from p to t (or an arrow $t\Box\!\!\longrightarrow\!\bigcirc\, p$ from t to p);
- a transition t is *activated* if
  1) for every place p from the pre-set of t the weight of the arrow from p to t is not greater than the number of tokens indicated at p,
  2) for every place p in the post-set of t the number of tokens at p increased by the weight of the arrow from t to p is not greater than the capacity of p;
- an activated transition t will *occur* in that the number of tokens at every place p is decreased by g if $p\bigcirc\xrightarrow{g}\Box\, t$ and in that the number of tokens at every place p' is increased by g' if $t\Box\xrightarrow{g'}\bigcirc\, p'$.

Formally, condition-event nets are the same as place-transition nets that have a capacity of *one* for each place and a weight of *one* for each arrow.

Based on what was said in Sect. 2.3 we can formulate the following definition:

Two transitions of a place-transition net are in conflict with one another if both are activated and the occurrence of one results in the deactivation of the other.

Fig. 35 illustrates a case of this kind.

Problem 13

Modify Fig. 30 such that
- *every producer places three objects in the channel on every delivery,*
- *every consumer takes two objects out of the channel on every receipt, at most one consumer is ready to receive in each case.*

Problem 14

Modify the system of read- and write-authorized processes in Fig. 33 such that each of the four processes can both read and write. In the non-active situation a process decides what it wants to do next, i.e. read or write. As in Fig. 33 a process can only write as long as no other process is reading.

3.4   Contacts and Their Avoidance Through Complementation

In accordance with what was said in Sect. 2.4, we can define the existence of a contact situation:

A contact exists under a marking M at a transition t if the places p of the form $p\bigcirc \xrightarrow{g} \square t$ contain at least g tokens but a place p' of the form $t\square \xrightarrow{g'} \bigcirc p'$ exists such that the number of tokens at p', increased by g', is greater than the capacity of p'. In other words, if t cannot occur due to insufficient capacity of a place.

As in the case of condition-event nets, contacts can be avoided through the construction of complements. If the capacity of a place p is not infinite, a place p' with "reversed arrows" is constructed in the manner indicated in Fig. 19 (see Fig. 36).

If p is a place with finite capacity in a place-transition net, a new place p' is constructed as a complement by

- adding a new arrow of the form $t\bigcirc \xrightarrow{g} \square p'$ with the same weight to every arrow of the form $p\square \xrightarrow{g} \bigcirc t$, and

- a new arrow of the form $p'\square \xrightarrow{g} \bigcirc t$ with the same weight to every arrow of the form $t\bigcirc \xrightarrow{g} \square p$.

- The capacity of p' is equal to the capacity of p and

- the initial marking of p' is equal to the capacity of p' (and, as such, p) minus the initial marking of p.

▍ Every place-transition net can be made contact-free by means of complements.

▍ The construction of complements does not change the ability of transitions to occur.

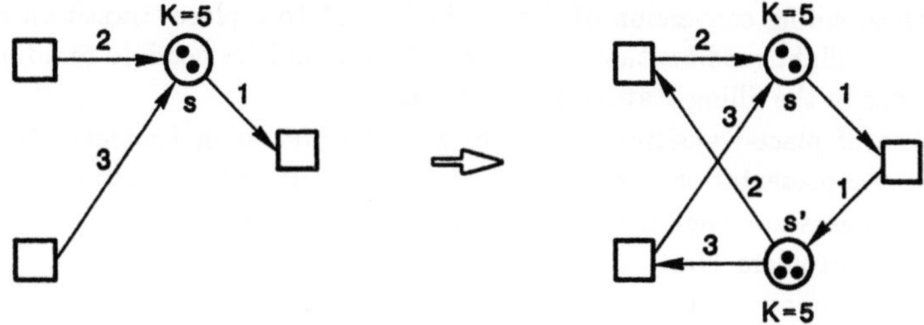

Fig. 36 Construction of a place complement

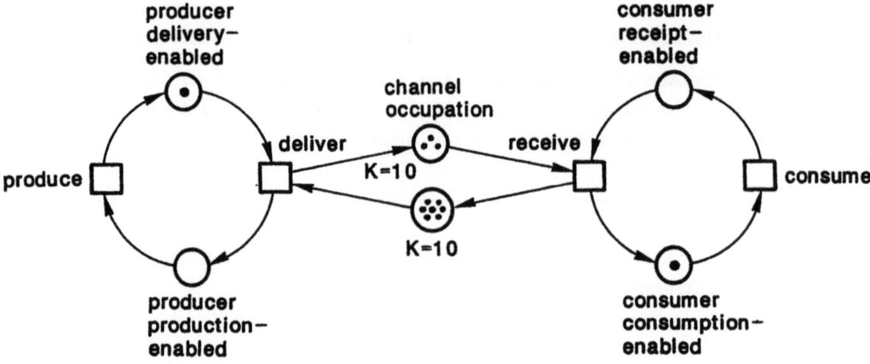

Fig. 37   Addition of the channel complement to Fig. 27

Problem 15

a) *In the net derived from Problem 13 construct a complement for every place (if necessary make use of appropriate capacities, i.e. capacities that do not change the behavior of this system).*

b) *Which of the complements constructed in a) are necessary in order to make the net contact-free and which are not?*

3.5   Further Examples

Having the possibility of storing more than one token in a place we can reinterpret Fig. 24 by allowing for the presence of several orders in the corresponding places. We can replace the labels 'order received', 'order processed by M_1' and 'order completed' with 'orders received', 'orders processed by M_1' or 'orders completed'. A capacity can be indicated for these places or it can be left undefined (in a real system corresponding to the diagram in Fig. 24 the capacity of these places is of course always restricted). It can be seen that even if we interpret Fig. 24 as a place-transition net, the places not discussed above will never contain more than one token.

A corresponding conversion of the net in Fig. 25 to a place- transition net only brings about slight modifications. In this case it would be possible to admit more than one car in the filling station and exit areas.

The use of place-transition nets is of greater interest in Problem 11. If only *quantitative* representations are important, i.e. if it is only important to indicate how many spaces are available and not which spaces are available, Fig. 38 provides a solution to Problem 11a.

A corresponding solution to Problem 11b would differ from the net indicated in Fig. 38 only in the initial marking, i.e. 'free spaces' and 'free cashiers' would each be assigned two tokens.

A combination of Problems 11a and 11b (i.e. two parking spaces per pump and two cashiers) would result in a complicated condition-event net. It would, for instance, contain eight different possibilities for paying. In Fig. 38 this combined problem can be solved with the help of place-transition nets by assigning a further token to 'free cashiers'.

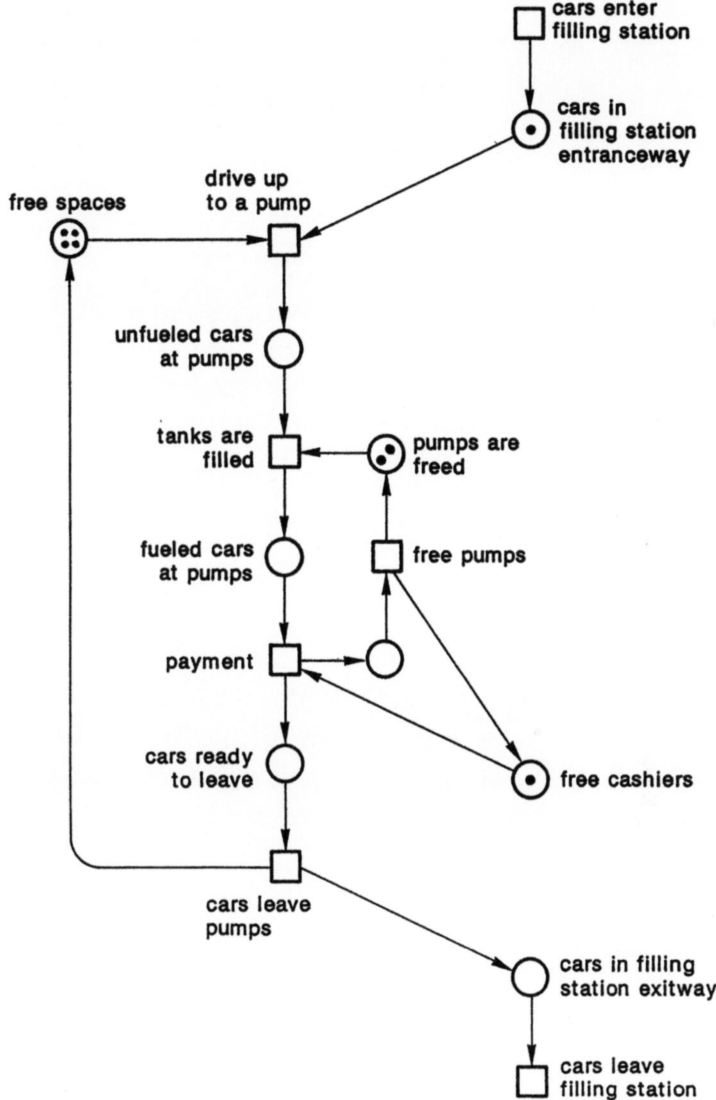

Fig. 38   Filling station with two pumps and two parking spaces per pump
represented as a place-transition net

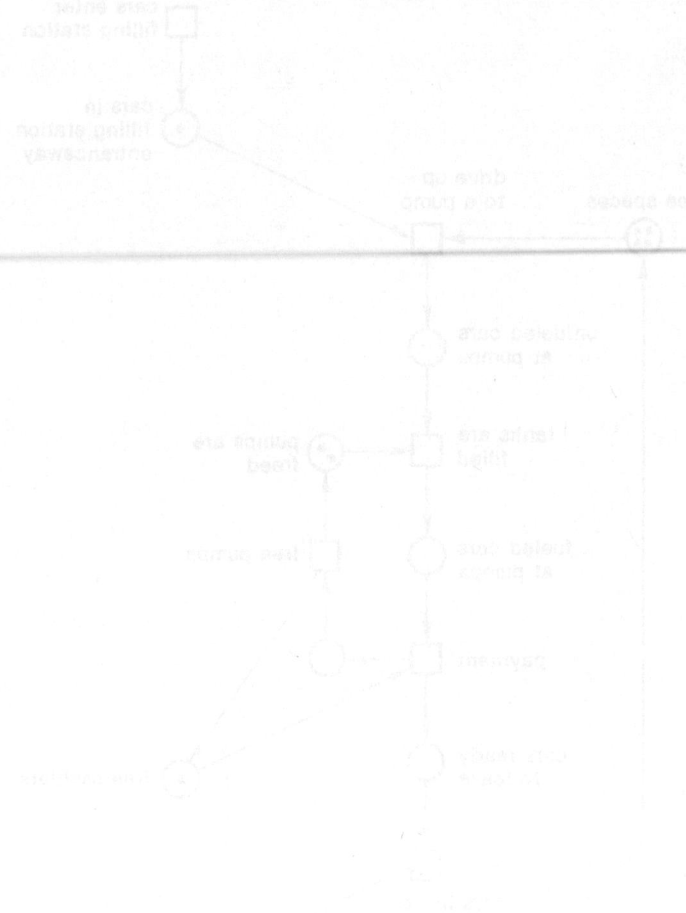

4   Individual-Token Nets

4.1   An Example of Constant Arrow Labels

Let us imagine a market. Dealers offer goods for sale and customers seek to buy the goods they need. When a dealer and a customer have agreed on a product and a price there is an exchange of goods and money. A business transaction takes place. After that, the dealer can acquire new goods and the customer new money and further business transactions will be possible.

A configuration of this kind is represented in Fig. 39 as a condition-event net involving a dealer C and two customers, A and B. This representation shows which customer the dealer sells to. However, it does not show what goods are involved, how expensive they are, etc.

If more than one dealer and two customers are involved in our market, its representation as a condition-event net will rapidly become complicated. As such, the use of place-transition nets was proposed in Chap. 3. Fig. 40 shows a corresponding place- transition net. However, it does not show *which* customer the dealer sells to. The representation only shows *that* a business transaction has taken place.

We would now like to combine the advantages of both types of nets and introduce a representation that
– indicates precisely who carries out a business transaction with whom and
– is still compact and easy to understand.

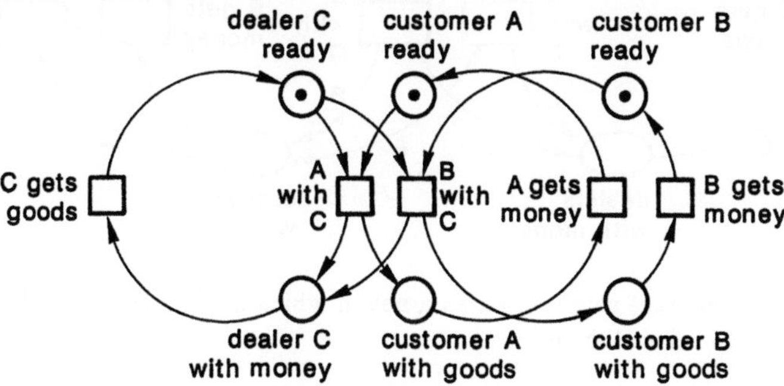

Fig. 39   A market represented as a condition-event net

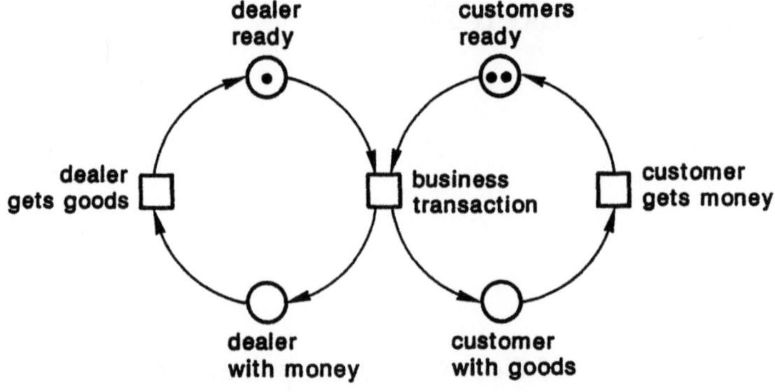

Fig. 40   The market represented as a place-transition net

Thus, we will no longer represent dealers and customers as indistinguishable tokens. Instead, we will show them as *tokens themselves*. In Fig. 41 the dealer C is, himself, the token of 'ready dealers'. The customers A and B are tokens of 'ready customers'. If A and C conduct a business transaction, this will result in the situation represented in Fig. 42. Here the transition 'A with C' has occurred. The arrows (i.e. the arrows that end or begin at the transition) are labelled with the letters 'A' and 'C'.

In Fig. 41 the rule applies that a token can only 'flow through an arrow' if this token corresponds to the arrow label. The fact that different arrows of a transition have the same label indicates where the tokens taken from 'ready dealers' or 'ready customers' will flow.

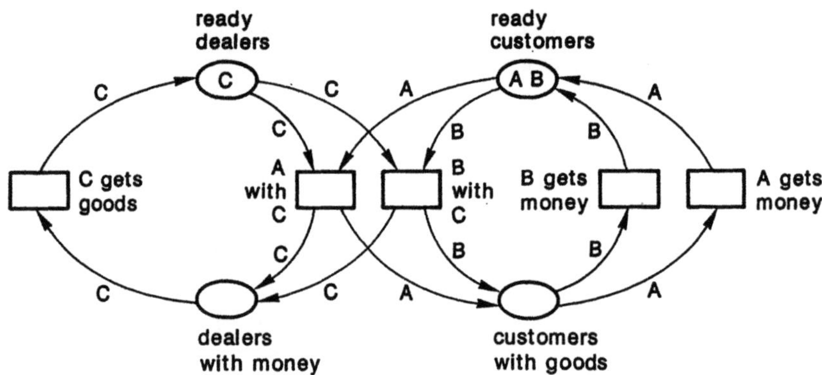

Fig. 41   The market represented as a net in which dealers and customers are them selves tokens

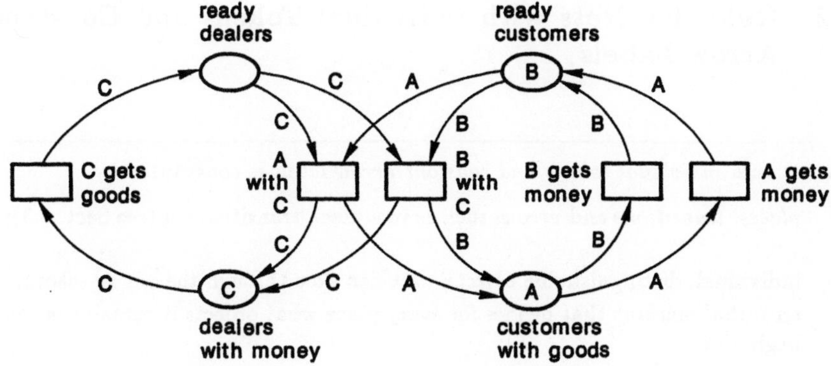

Fig. 42   Situation following the occurrence of transition "*A* with *C*" in Fig. 41

Fig. 43 illustrates once again the principle on which the occurrence of a transition is based in a net with individual tokens and constant arrow labels.

In order to return to the situation in Fig. 41 from the configuration represented in Fig. 42, the two transitions '*C* gets goods' and '*A* gets money' have to occur.

In summary we can say that in Fig. 41 and 42 we succeeded in representing what is represented in Fig. 39 without a loss of information. It is precisely indicated what customers the dealer transacts business with and who acquires goods or money. There are fewer places than in Fig. 39. The number of transitions is the same. If we add further dealers and customers to the market we get further transitions in Fig. 41 but no additional places. In Sect. 4.4 we will see how we can also reduce the number of transitions.

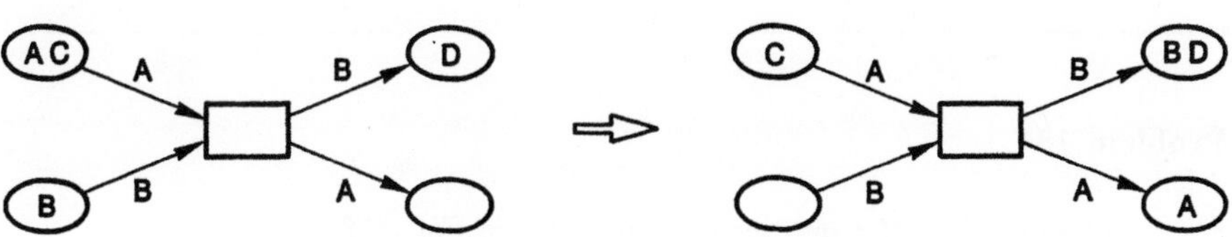

Fig. 43   Occurrence of a transition in a net with constant arrow labels

4.2 Rules for Nets with Individual Tokens and Constant Arrow Labels

A *net with individual tokens and constant arrow labels* is constituted by

- *places, transitions* and *arrows* such as in a place-transition net (see Sect. 3.3);

- individual, distinguishable *objects*, that can flow through the net as tokens;
- an *initial marking* that defines for every place what objects it contains in the beginning;
- a *label* on every arrow designating an individual object.

The pre- and post-set of a place or transition is explained as in the case of place-transition nets.

In a net with individual tokens and constant arrow labels

- a *configuration* is constituted by a distribution of objects to the places;
- a transition t is *activated* if every place p in the pre-set of t contains the object designated by the label of the arrow from p to t;
- an activated transition t *occurs* in that
  1) the object is removed from every place p in the pre-set of t that is indicated by the arrow from p to t, and

  2) every place p' in the post-set of t contains the object indicated by the arrow from t to p'.

Problem 16

a) How many different markings are attainable in Fig. 41?

b) Modify Fig. 41 such that two dealers are present. Each dealer can conduct business transactions with each customer.

4.3 Further Possibilities for Constant Arrow Labels

In Sect. 4.1 it was shown how the objects that can be stored as tokens in places can also appear as arrow labels. It is possible for objects to appear 'out of the blue' or to vanish 'without a trace'. We can take the producer-consumer system we used in chapters one and two as an example (see Fig. 11).

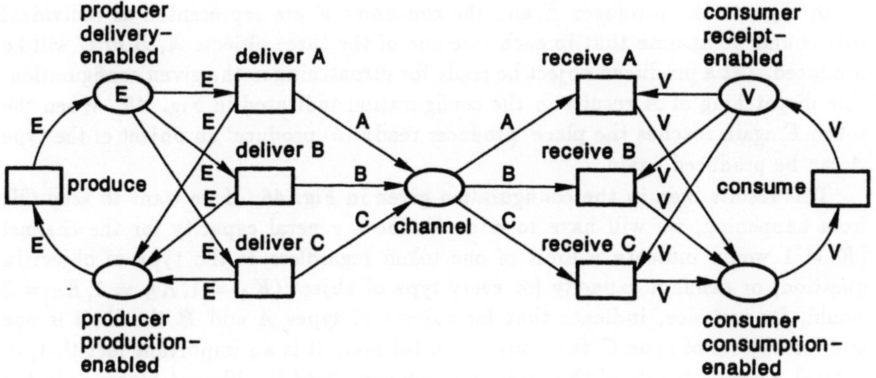

Fig. 44 The producer-consumer system with individual tokens

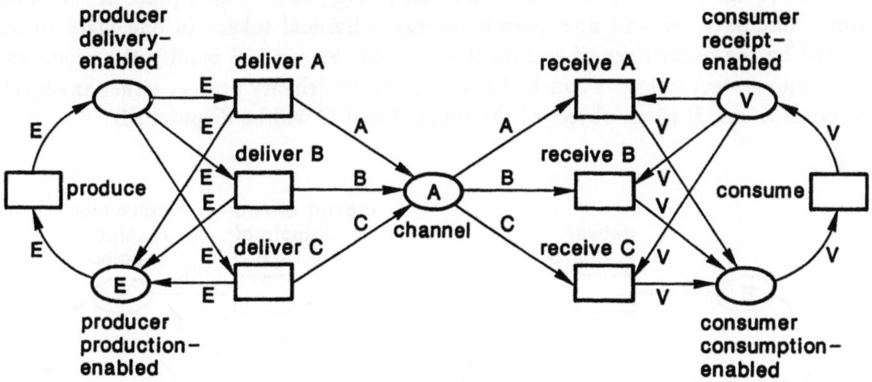

Fig. 45 Configuration following production of object A

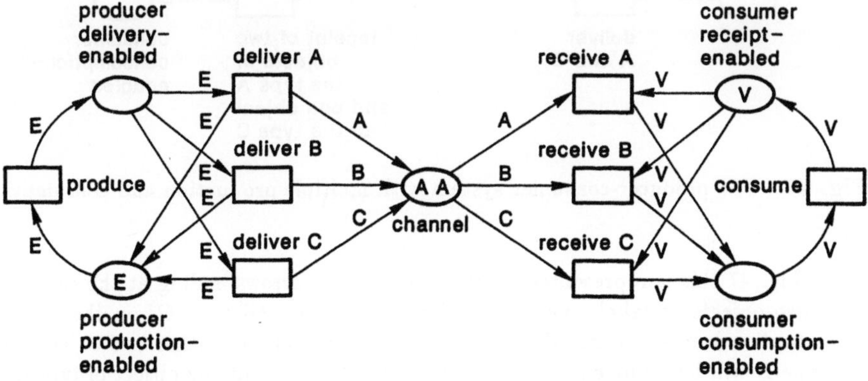

Fig. 46 A second object of the type A has been produced

In Fig. 44 the producer E and the consumer V are represented as individual tokens and we assume that in each case one of the three objects A, B or C will be produced. Let a produced object be ready for dispatching in the given configuration. The dispatching of A results in the configuration indicated in Fig. 45. When the token E again reaches the place 'producer ready to produce' an object of the type A can be produced again.

This results then in the configuration given in Fig. 46. If we want to keep this from happening, we will have to either define a general capacity for the channel ($K = 1$ would indicate a limit of one token regardless of the type of object in question) or define a capacity for every type of object ($K_A = 1, K_B = 1, K_C = 2$ would, for instance, indicate that for objects of types A and B the limit is one and for objects of type C the limit is two tokens). It is an important fact that, in general, several objects of the same type are permitted in different places or in the same place.

The switch from conditions and events to places and transitions has made it possible to add or remove several 'black' tokens ($\odot$) to or from a place at the same time. Similarly, we will now permit several individual tokens in one place to be moved by the occurrence of a transition. To do this we will modify the producer-consumer system in Fig. 44 such that now in every delivery process either an object of types A and B or an object of the types A and C will be dispatched.

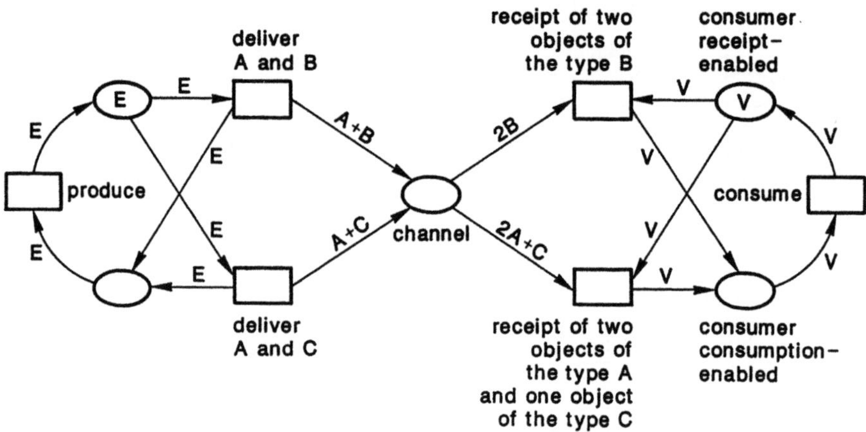

Fig. 47   The producer-consumer system with multiple production and consumption

In Fig. 47 this is represented by the fact that the arrows ending at the channel have two objects as labels, i.e. '$A + B$' on one arrow and '$A + C$' on the other. Consumption is now organized such that in any removal step either two objects of the type B will be removed or two objects of the type A and one object of type C. This is represented by the labels '$2B$' or '$2A + C$'.

We want to modify the configuration in Fig. 44 such that

1. only objects of type A are produced,

2. two consumers V_1 and V_2 are present and

3. the producer defines which of the two consumers will receive the object produced. A token can now also be a pair (A, V_1) or (A, V_2). Fig. 48 shows the corresponding system with a configuration in which consumer V_1 can consume the object A but the consumer V_2 cannot.

In nets with individual tokens it is sometimes advantageous if a place p is located both in the pre- and post-sets of a transition t, i.e.

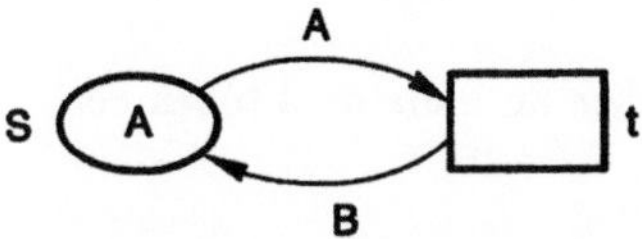

when t occurs A is removed from the place p and B is added. A configuration of this kind is called a *loop*. An example in which it can be applied is the system of the four seasons in Fig. 14. In Fig. 49 this system is represented as an individual-token net. Each of the four seasons appears as a token.

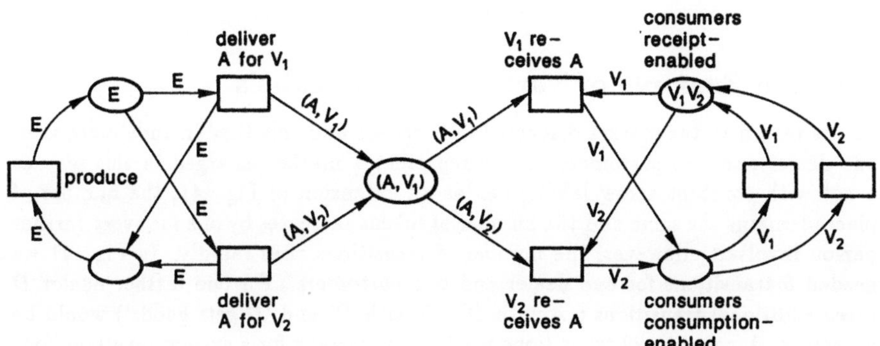

Fig. 48   The producer-consumer system with pre-specified consumers

Problem 17
Add to Fig. 49 such that the condition represented in Fig. 14, i.e. ' it is winter or spring', can be seen.

Problem 18
Solve Problem 12 with a net in which the ferryman, goat, wolf and cabbage are individual tokens. The net should contain as few places as possible.

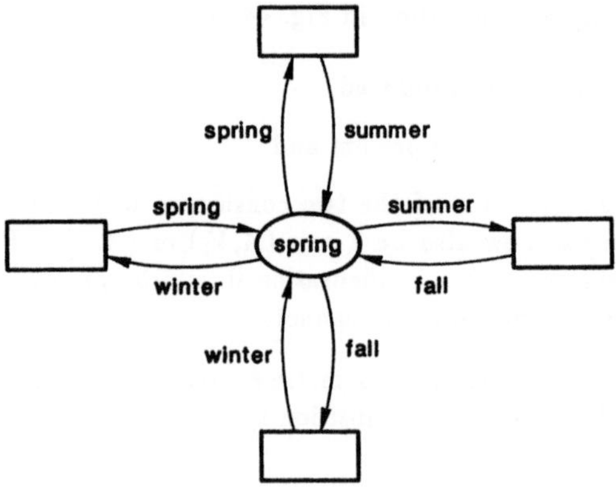

Fig. 49   The system of the four seasons represented as an individual token net

Problem 19
Convert Fig. 24 to an individual-token net with as few places as possible.

Problem 20
Convert Fig. 25 to an individual-token net with as few places as possible.

4.4   An Example of Variable Arrow Labels

Let us return to the market discussed in Sect. 4.1 and this time include more than
one dealer and two customers. If we represent a market enlarged in this way as
a net with constant arrow labels, i.e. as an extension of Fig. 41, the number of
places remains the same and the number of tokens increases by one for every further
person involved. However, the number of transitions rises rapidly. In Fig. 41 we
needed 5 transitions for one dealer and two customers. For one further dealer D
three additional transitions ('A with D', 'B with D' and 'D gets goods') would be
necessary. A net with 29 transitions would be necessary for a system involving four
dealers and five customers.

Thus, we need to find a representational technique that does not involve a change
in the underlying net when dealers or customers are added.

A system of this kind is possible based on the observation that the transitions
'A gets money' and 'B gets money' in Fig. 41 have the same pre- and post-sets.
Both transitions take a buyer A or B into a 'purchase-ready' state. The occurrence
of these transitions indicates that a customer x gets money, in which case $x = A$ or
$x = B$. Based on this idea, we can now devise a transition 'x gets money' (Fig. 50)
and have this transition occur *with respect to one of the customers A or B*. This
means that prior to occurrence the variable x is replaced by A or B on the arrows
that begin or end at 'x gets money'. Afterwards, it occurs as in a net with constant

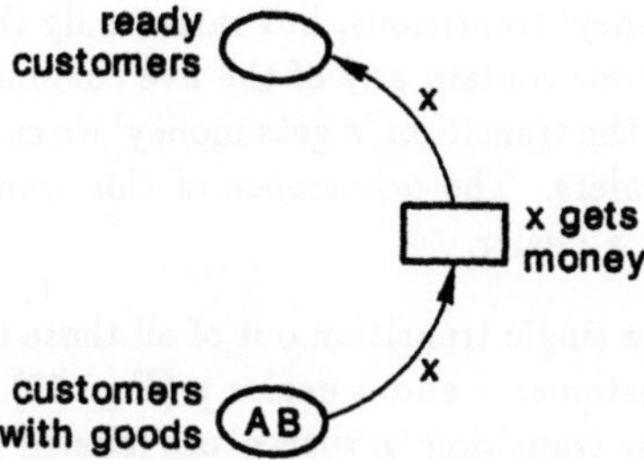

Fig. 50   Construction of a transition with the variable x

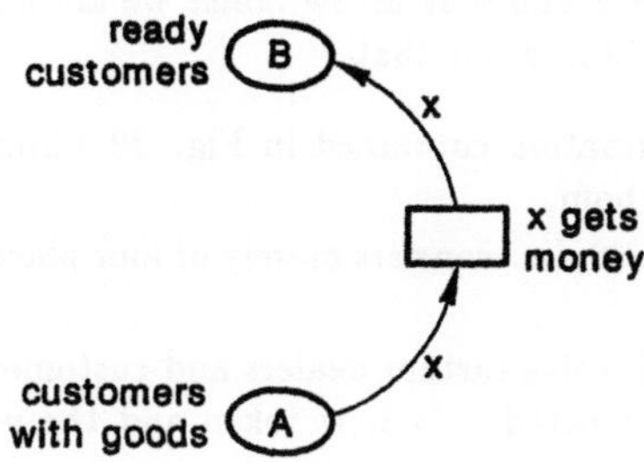

Fig. 51   Situation following the occurrence in Fig. 50 of the transition "x gets money" with respect to B

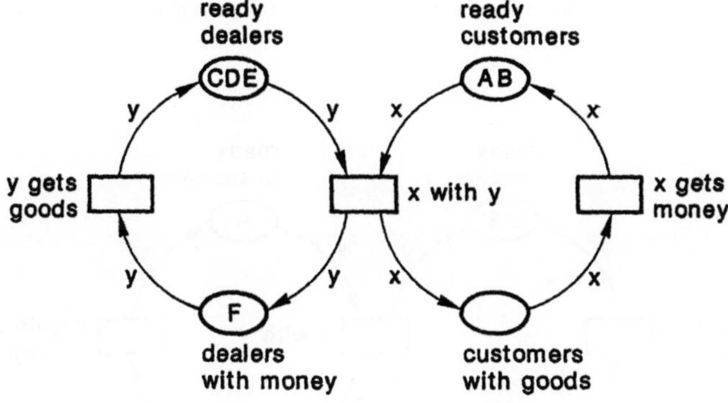

Fig. 52   The market with two customers and four dealers

arrow labels. Thus, if x is replaced by A, the transition "x gets money" occurs like the transition 'A gets money'. On the other hand, if x is replaced by B the transition 'x gets money' occurs like the transition 'B gets money'. This change is shown in Fig. 51.

This technique offers a clear advantage. If further customers are involved in the market, e.g. if there are another three in addition to A and B, there is no need to represent five '... gets money' transitions, but rather only the one shown in Fig. 50. The places involved can now contain any of the five customers as a token.

Just as we introduced the transition 'x gets money' we can also devise a transition 'y gets goods' for the dealers. The occurrence of this transition requires that the variable y be replaced by a dealer.

Finally, we can make a single transition out of all those that represent a business transaction between a customer x and a dealer y (Fig. 52).

Here, the arrows of the transition 'x with y' are labeled with two variables x and y. For this transition to occur, x must be replaced by a customer and y by a dealer. Thus, the transition "x with y" in Fig. 52 can occur when x is replaced by A or B and y by C, D or E. Fig. 53 shows this transition when x is replaced by B and y by D. With the variables x and y as arrow labels we are now able to represent the market depicted in Fig. 39 in a net that

- contains all the information contained in Fig. 39 while visibly indicating who does business with whom,
- is as compact as Fig. 40, i.e. consists merely of four places and three transitions, and
- makes it possible to involve further dealers and customers in the market in that every new person is treated as a new token and the net does not need to be changed.

As a further example let us consider a market for housing rentals. Here, a third group of persons is involved in addition to landlords and tenants, i.e. agents. When a rental contract is concluded, the involvement of an agent is possible but not necessary. Fig. 54 shows the organization required for three tenants, three landlords and two agents.

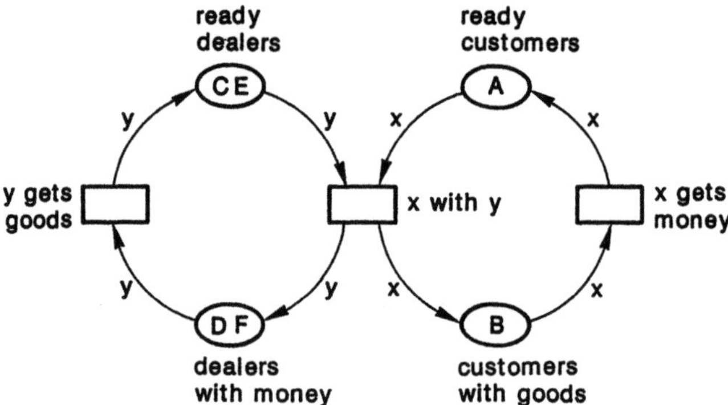

Fig. 53   Situation following the occurrence in Fig. 52 of the transition "x with y" with respect to $x = A$ and $y = D$

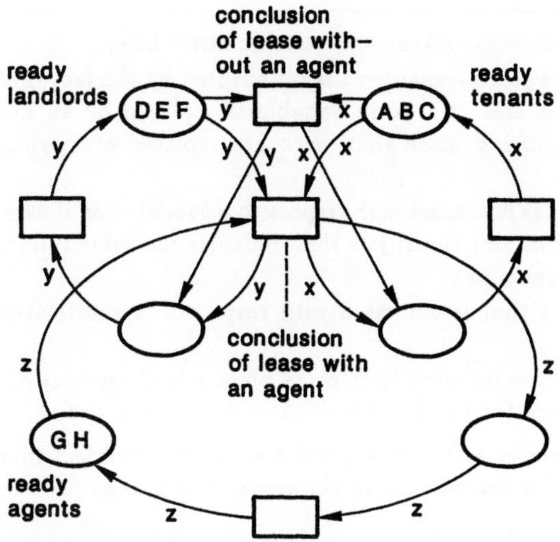

Fig. 54   Housing rentals market with the possible participation of brokers

4.5   Rules for Nets with Individual Tokens and Variable Arrow Labels

A *net with individual tokens and variable arrow labels* is constituted by
- *places, transitions, arrows* and an *initial configuration* consisting of individual objects such as required in Sect. 4.2 for nets with constant arrow labels;
- a variable, e.g. x, y or z or something similar as a label on every arrow.

As in the case of constant arrow labels, the pre- and post-sets of a place or transition is explained and a configuration is given by a distribution of objects among the places.

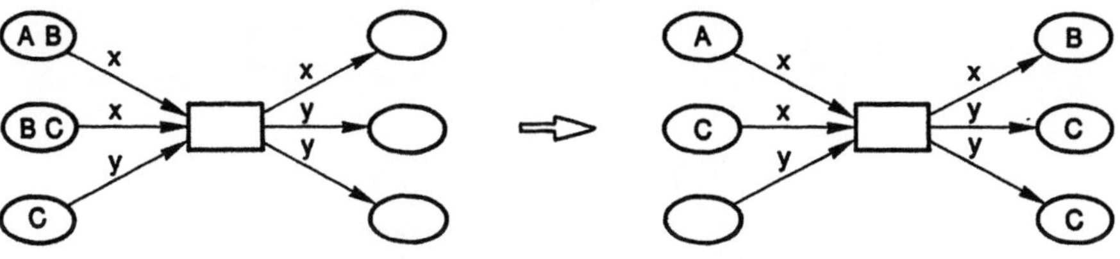

Fig. 55   Occurrence with variable arrow labels

In a net with individual tokens and variable arrow labels,

- a *substitution for a transition t* is constituted by the fact that on the arrows that begin or end at t every variable is replaced by an individual object; variables occurring more than once are replaced everywhere by the same object;

- a transition t is activated with respect to a substitution if every place p in the pre-set of t contains the object that indicates the substitution of variables on the arrow from p to t;

- a transition t that is activated with respect to a substitution will occur in that

  1) the object is removed from every place p in the pre-set of t that has been substituted for the variable of the arrow from p to t,

  2) every place p' in the post-set of t acquires the object that has been substituted for the variable of the arrow from t to p'.

Fig. 55 provides another example of the way in which transitions occur with variable arrow labels. In the configuration given, t can occur with $x = B$ and $y = C$. No other possibility exists.

In the examples given in Sect. 4.4 the arrow labels are always organized such that an object (e.g. customer A) can only replace a specific variable. This is not always possible. As an example let us consider a market at which each of the persons involved takes turns in assuming the role of a dealer and a customer. The last person to buy an object sells it again. The last person to sell something buys something new. Fig. 56 shows this system for four persons. In this configuration t_0 last occurred with $x = B$ and $y = D$. Now t_1 can occur with $y = B$ and t_2 with $x = D$. After that t_0 can occur again, this time with $x = D$ and $y = B$.

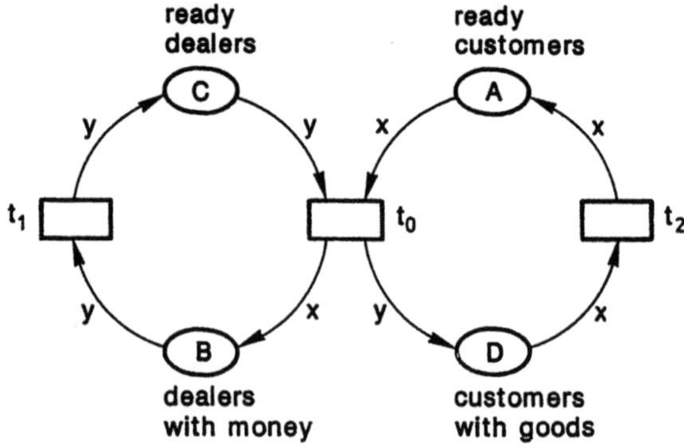

Fig. 56   The persons involved appear alternately as dealers and customers

4.6   Further Possibilities for Variable Arrow Labels

In Sect. 4.3 we saw that constant arrow labels can be grouped together. Several objects can be placed on an arrow (Fig. 47) which jointly flow 'through the arrow' when the transition in question occurs. In Fig. 48 objects are grouped together in pairs and, in this way, are combined to form single tokens. Similar structures are also possible with other labels.

As in Sect. 4.3 we will consider, first of all, objects that appear 'out of the blue' or vanish 'without a trace'. The second case causes no difficulty. In p ◯—x→▢ t one of the objects in p disappears when t occurs if no arrow from t bears the label x. Similarly, in t ▢—x→◯ p an arbitrary object is produced and stored in p. In general, the intention is not to produce a fully arbitrary object, but rather one that has certain characteristics. These characteristics can be specified in *transition* t.

As an example, let us consider once again the producer-consumer system in Fig. 11, this time for use in the transmission of messages. In this case, 'deliver' becomes 'transmit' and 'receive' remains 'receive'. It is important that the contents of the message be taken into account. We will assume that four-bit messages (i.e. information consisting of four '0' or '1' digits) are transmitted. Fig. 57 shows a net of this kind. The transition 'transmit' can occur with $x = A$ and an arbitrary four-bit message for y. The configuration given in Fig. 58 is produced with $y =$ 0110. As a further addition to Fig. 11 there is a component that stores all the messages received.

Thus, occurrence of the transition 'receive' with $y = 0110$ and $z = B$ in Fig. 58 results in the situation indicated in Fig. 59.

The variables in Fig. 57 have been selected such that x is always replaced by A, z by B and y by a four-bit message. Needless to say, this is not necessary.

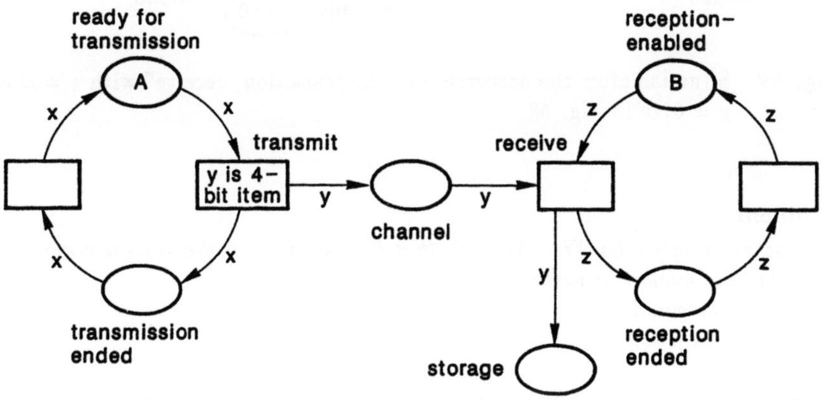

Fig. 57   Transmission and counting of messages dependent on their contents

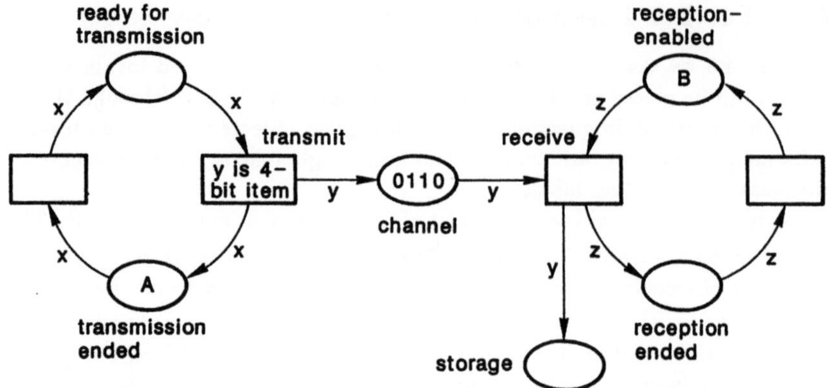

Fig. 58   Situation following the occurrence of the transition "transmit" with $x = A$ and $y = 0110$ in Fig. 57

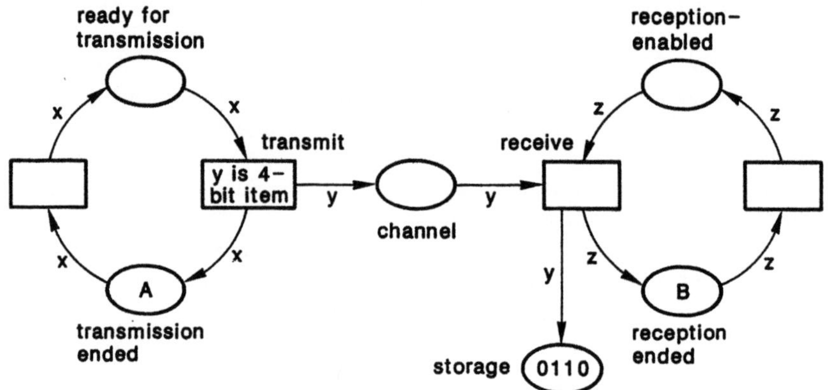

Fig. 59   Situation after the occurrence of the transition "receive" with $z = B$ and $y = 0110$ in Fig. 58

Problem 21

Choose other labels for Fig. 57 such that the behavior of the system remains the same but the number of variables used is
a) as large as possible,
b) as small as possible.

Fig. 60 shows arrow labels with more than one variable. When the transition 'receive' occurs, two four-bit messages will be removed from the channel and placed in storage in one step. These two messages can be any two messages in the channel whether the messages are the same or different. In Fig. 61, on the other hand,

48

'receive' can only occur when a message is present twice in the channel. In this case, both copies of the message are removed from the channel, but only one is placed in storage.

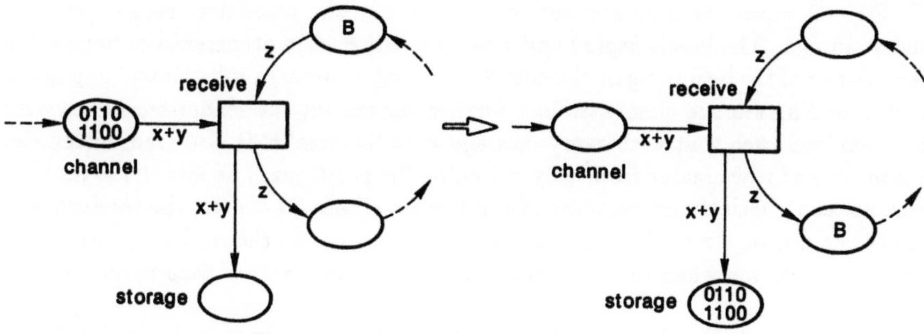

Fig. 60   Coincident reception of two messages

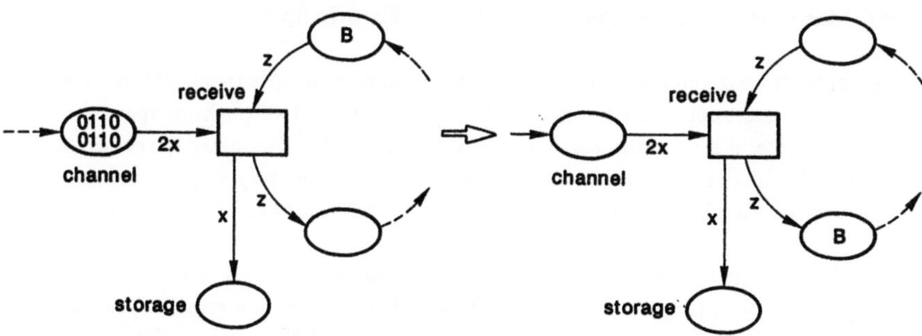

Fig. 61   Reception of double messages

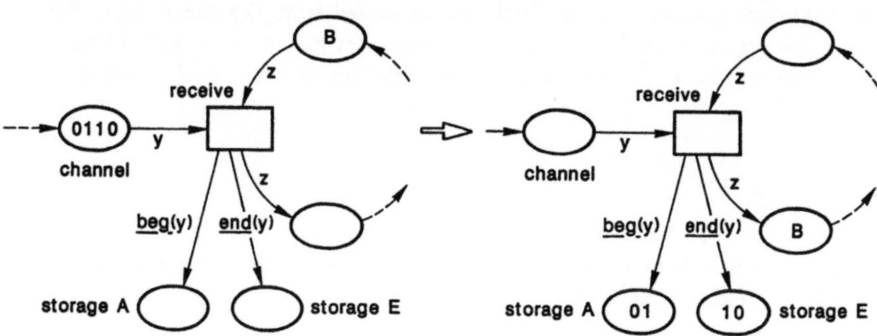

Fig. 62   Division of messages received into beginnings and ends

To acquaint ourselves with another kind of arrow label, let us modify the sender-receiver system in the following way: every four-bit message sent is broken down by the receiver into two parts of two digits each. The beginning of every word z, $beg(z)$, consists of the first two digits and the end, $end(z)$, of the last two. The beginnings and the ends of words received are stored separately in the storage elements A and E. Fig. 62 shows the modified net component, i.e. the transition 'receive' with its surroundings. The labels $beg(z)$ and $end(z)$ result on the occurrence of 'receive' in the storage of the beginning or the end of the four-bit message substituted for y in the corresponding storage elements. In a further variant let the sender-receiver system be organized such that with every message x its 'inverse' x' is also transmitted (let the inverse of x be created from x by switching the positions of '0' and '1') so that the receiver can check the correctness of the transmission. Of course, the receiver only stores the message x itself in the storage element. Fig. 63 shows the organizational structure involved when the transitions 'transmit' and 'receive' each occur once.

In Fig. 48 it was shown that pairs of objects can appear as arrow labels. A similar construction is also possible with pairs of variables. As an example, let us organize the sender-receiver system such that two receivers, B and C, are given and the sender, by analogy with the producer in Fig. 48, determines the receiver (either B or C) for every message. Fig. 64 illustrates this principle.

Let us now look once again at the filling station example from Chaps. 1 and 2. In Fig. 38 we assumed the existence of two pumps with two parking spaces at each pump. The attainable configurations of Fig. 38 indicate how many spaces or pumps are available. In Fig. 65 we can indicate much more precisely what parking spaces and pumps are involved. L and R are the individual   tokens of the pumps.

We also use these symbols for the parking spaces next to them. Since the two parking spaces next to pump L are not distinguished from one another, two L's are used as tokens in the place 'available parking spaces'. The same applies for R. We define two assistants, T and U.

In one variant, let the filling station be organized such that four arbitrary parking spaces are available, all of which can be served from the two pumps. Fig. 66 shows this organizational diagram. In this case, not just pairs (x, y) but also three-variable groups (x, y, z) are used as labels. Needless to say, lists of this kind can be longer still.

Problem 22

Let five persons be involved in a market. After engaging in a business transaction, every person reverts to an idle state and decides whether he or she wants to assume the role of a buyer or seller the next time.

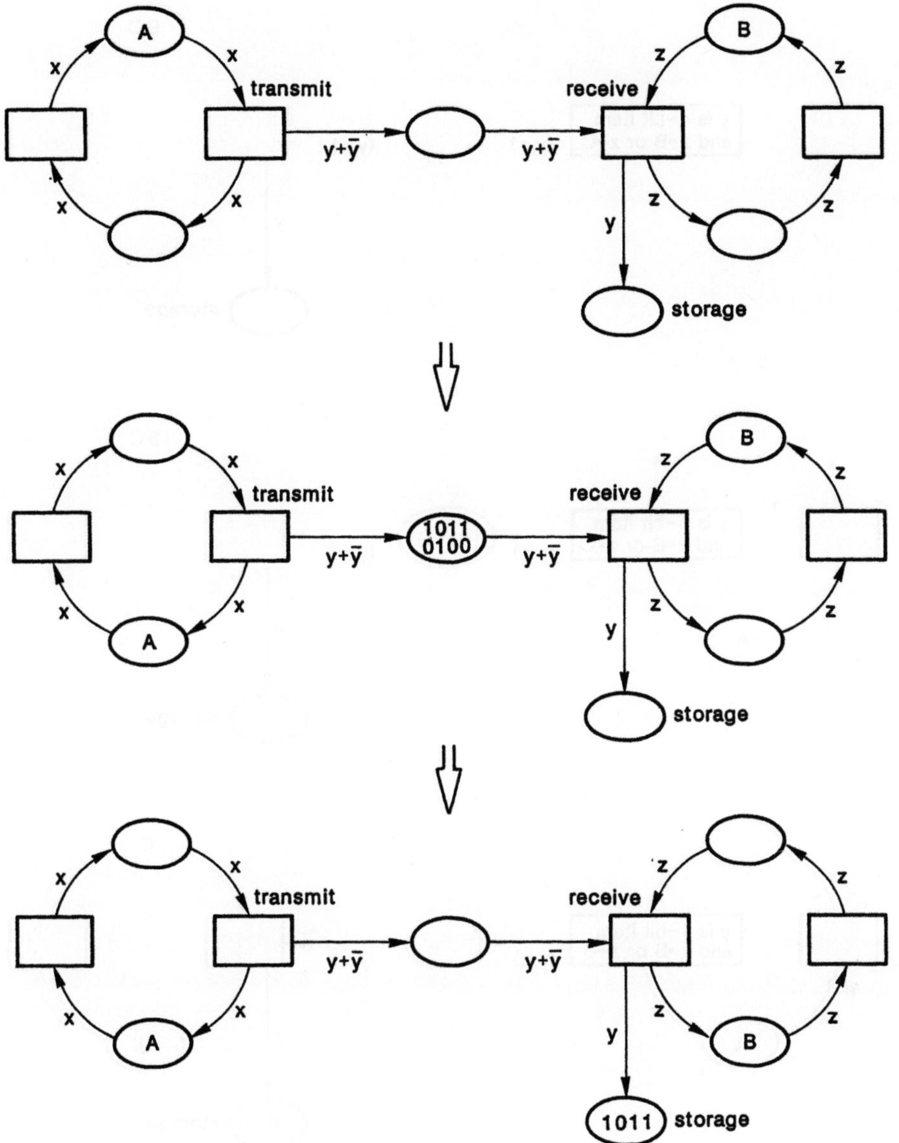

Fig. 63   Transmission of messages and their inverses

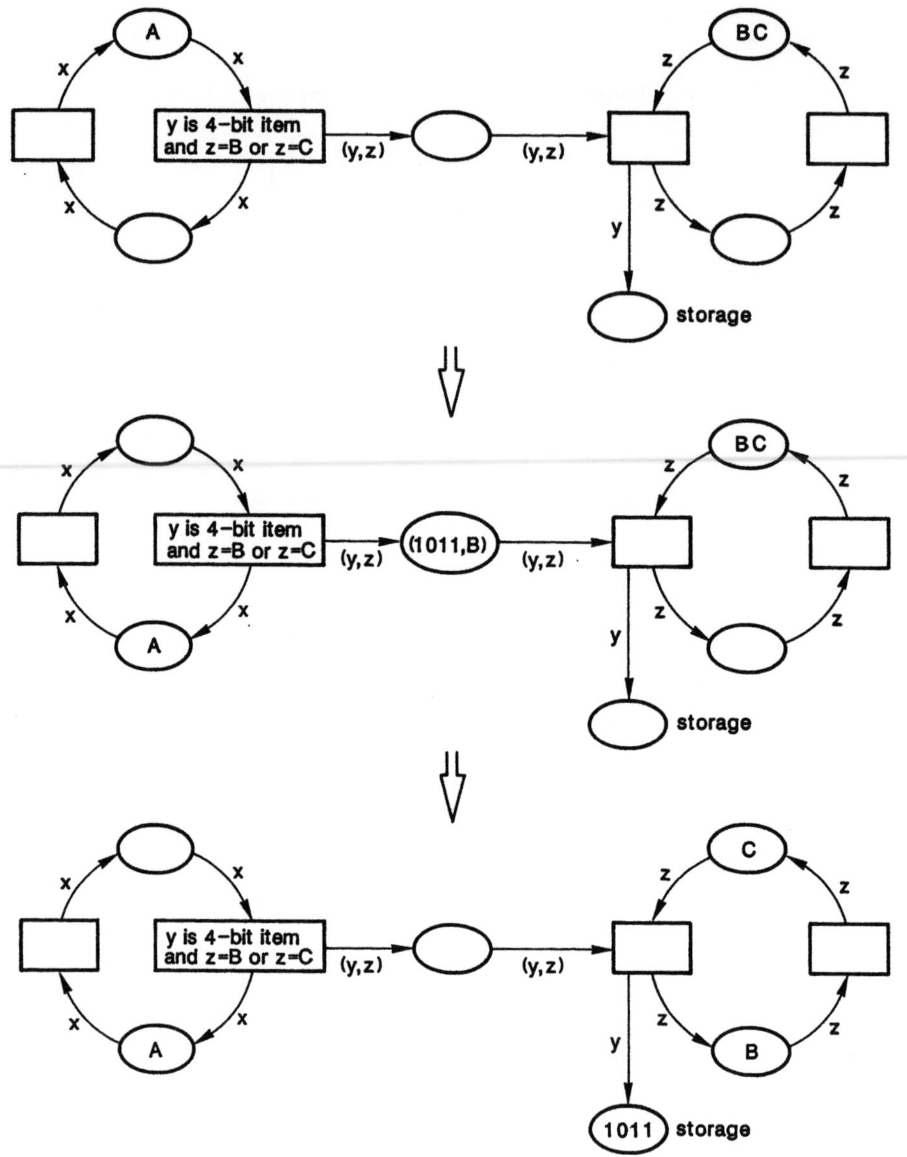

Fig. 64   Transmission of messages with specification of receiver

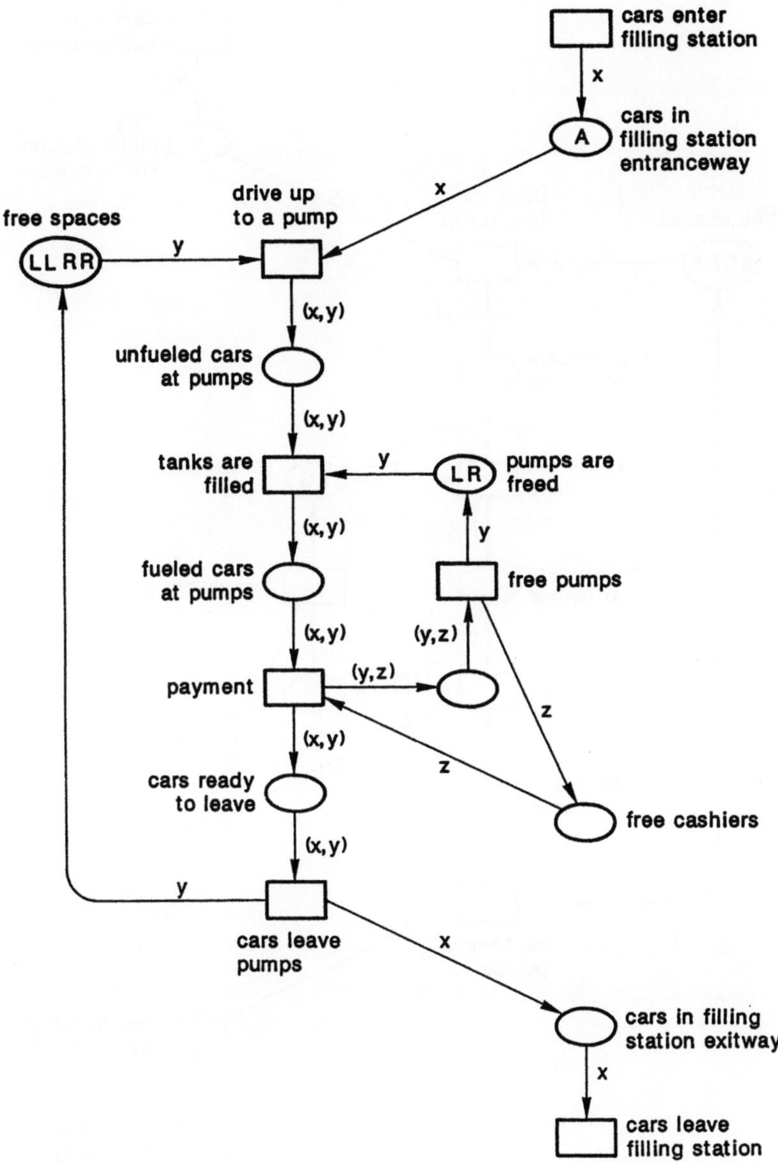

Fig. 65   Filling station with two attendants and two pumps with space for two cars per pump

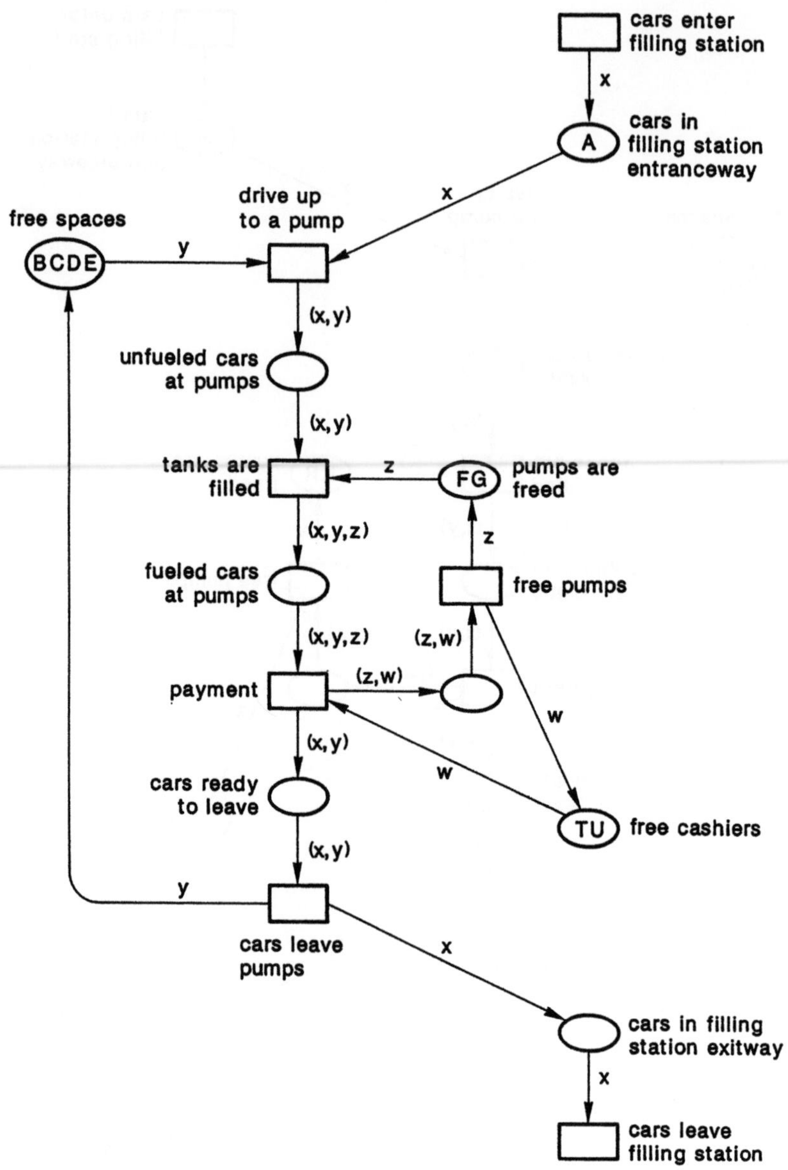

Fig. 66 Filling station with two cashiers and two pumps, both of which are accessible from four spaces

Problem 23

a) Transform Fig. 24 into an individual-token net with as few transitions as possible (see Problem 19).

b) Transform Fig. 49 into an individual-token net with as few places and transitions as possible.

Problem 24

Add a storehouse and a bank to the market in Fig. 52. Dealers pick up goods from the storehouse and take their money to the bank. Customers withdraw money from the bank and place purchased objects in the storehouse.

4.7 Individual-Token Nets

We come now to the most general form of individual-token nets. Nets of this kind can include arbitrary combinations of constants, variables and operations (such as $beg(y)$ in Fig. 58) as arrow labels. Every replacement of variables by objects for the entire label indicates what objects are meant and how often. (In a sender-receiver system, for instance, $2x + beg(y) + A$ could be a label.) The transitions can have condition-like inscriptions that have to be fulfilled on occurrence.

The 'black' token (⊙) used in the first two chapters plays a special role among the constant tokens. This kind of token label can be omitted on an arrow. Fig. 67 shows the addition to Fig. 57 of a 'channel available' place that can contain a maximum of one black token. Similarly, Fig. 68 shows a representation of the read- and write-authorized processes in Fig. 33. In this case, the processes can be individually distinguished. On the other hand, it would not be sensible to distinguish the keys.

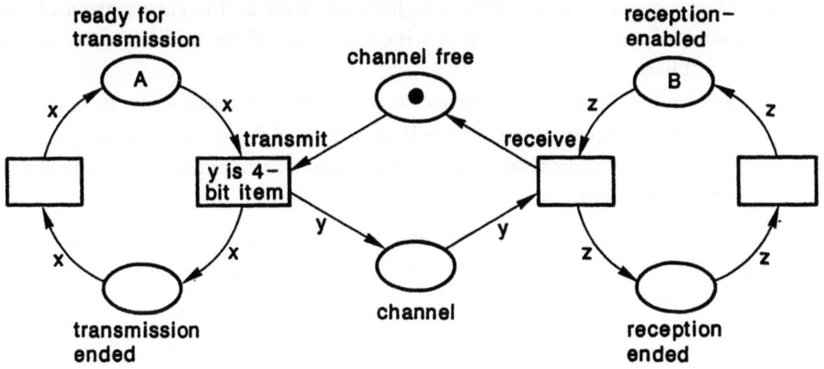

Fig. 67   Addition of "channel available" to Fig. 57

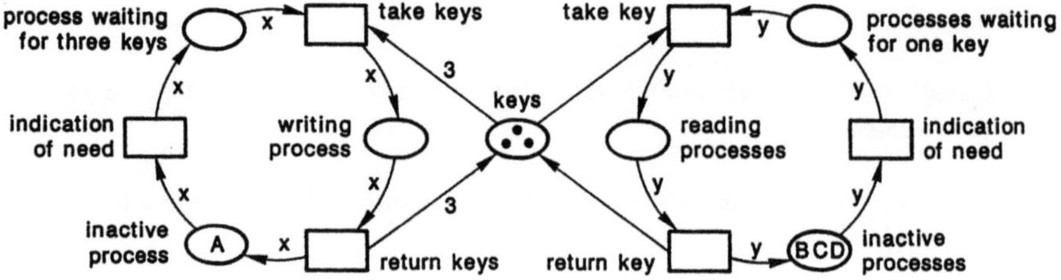

Fig. 68 Individual read- and write-authorized processes

An *individual-token net* is constituted by

- *places, transitions* and an *initial configuration* consisting of individual objects such as called for in Sects. 4.2 and 4.5 for nets with constant and/or variable arrow labels;

- an expression consisting of *constants, variables* and *operations* as a label for every arrow so that the substitution of variables with objects indicates a number of objects, it being possible for the same object to be indicated more than once;

- an *additional condition* for every transition (which can be omitted).

In an individual-token net

- a *substitution for a transition t* is given when every variable in the expressions labeling the arrows beginning or ending at t and in the additional condition of t is replaced by an individual object; variables occurring more than once are replaced by the same object;

- a transition t is *activated with respect to a substitution* if every place p in the pre-set of t contains every object at least as often as the substitution of the expression of the arrow from p to t indicates and if the additional condition in t is fulfilled;

- a transition t activated with respect to a substitution will *occur* in that

  1) the same number of objects will be removed from every place p in the pre-set of t as indicated by the replacement of the expression of the arrow from p to t, and

  2) every place p in the post-set of t has added to it the same number of objects as indicated by the substitution of the expression of the arrow from t to p.

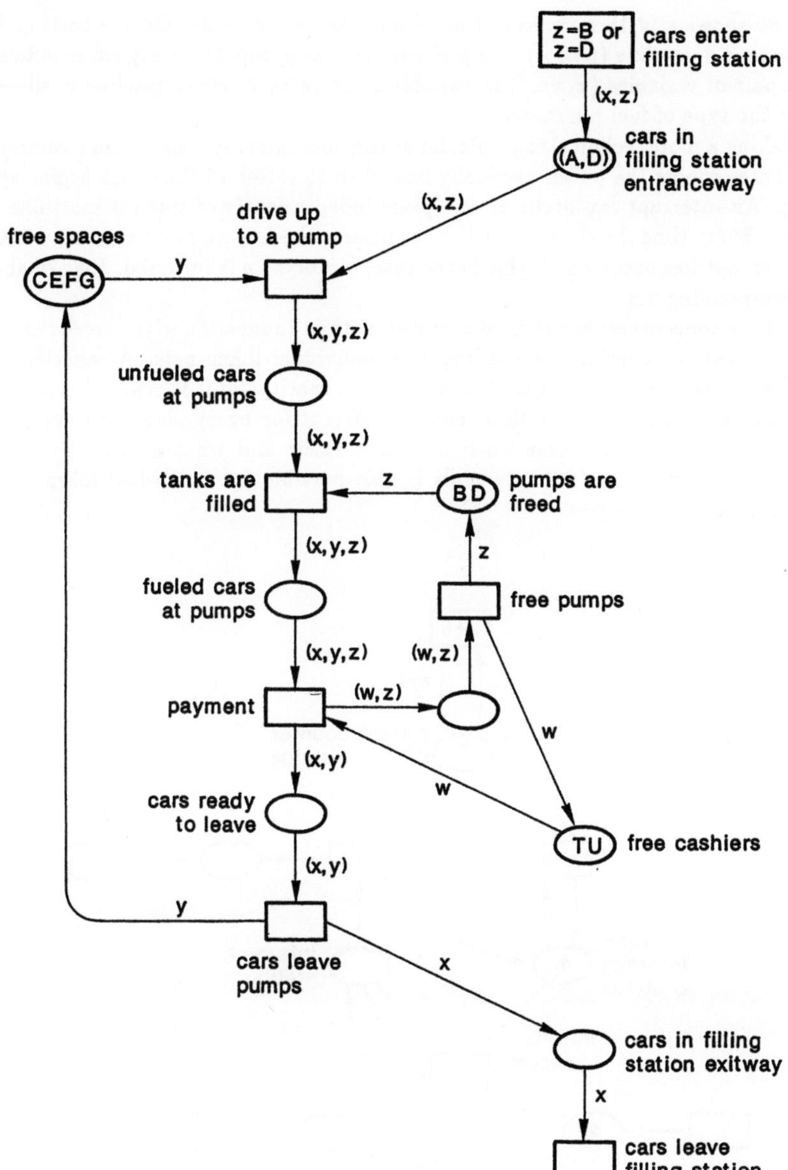

Fig. 69   Filling station with gasoline and diesel pumps

Fig. 69 shows a further variant of our filling station example. On the basis of Fig. 66 there is a gasoline (petrol) pump B and a diesel pump D. Every car is indicated by a pair of variables (x, z). The variable z indicates whether gasoline or diesel is to be the type of fuel purchased.

Taking a more technical example, let us consider interrupt control on a computer. A counter counts the pulses cyclically from 0 to 15 (after 15 the count begins again at 0). An interrupt can occur at any place independently of this (at most one per cycle). Every time the clock emits the 9th pulse a check is carried out to see whether an interrupt has occurred. If this is the case, a process p is initiated. Fig. 70 shows a corresponding net.

All the concepts we acquainted ourselves with in connection with place-transition nets can now be developed accordingly for individual-token nets. A capacity limit can be defined for places such that when the capacity limit is reached transitions cannot occur. The capacity limit can be different for every place and every type of object. It should be clear when a conflict exists and what a contact situation is. The construction of complements is also possible for individual-token nets if capacity limits are given.

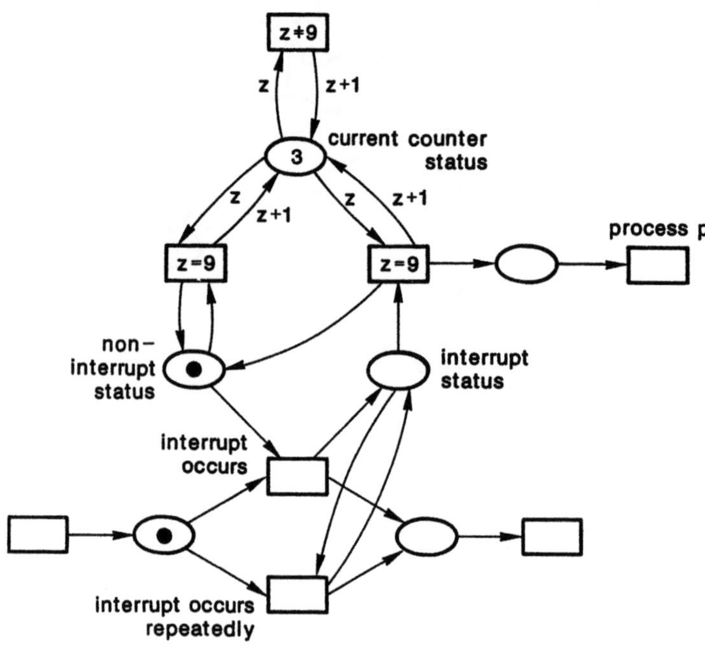

Fig. 70   A simple interrupt control

Problem 25

Make the individual processes visible in the net developed in connection with Problem 24.

Problem 26

In a variant of Fig. 69 let there be three parking spaces. In the first only gasoline is sold, in the second only diesel and in the third both.

Problem 27

Add appropriate arrow labels to Fig. 6 or Fig. 10, creating an individual-token net that models the behavior described in Sect. 1.1.

5   Channel-Agency Nets

Thus far, we have dealt with nets that represent the dynamic behavior of real systems. A large number of different factors have to be taken into consideration when designing such nets, e.g. what components are involved, how they behave in every possible situation, how tokens are to be distributed initially, whether or not all dependencies are correctly represented, etc.

A 'one-step' design of such nets is complicated and likely to produce errors if the net in question is bigger than in the examples dealt with thus far. In such cases it is a good idea to start out with an initial design making use of ordinary-language inscriptions. There is, of course, no precise occurrence rule for nets of this kind, which are referred to as *channel-agency nets*.

5.1   Examples

In the initial approach to the modelling of a real system it is a good idea to break the system down into a small number of discrete parts, the separation of which appears to be useful. Fig. 71 shows an initial rough breakdown of a producer-consumer system. It consists basically of a producer, a transmission channel and a consumer. The producer and the consumer are not directly connected, but only by means of the channel.

Fig. 72 shows an initial structure of a system consisting of reading and writing processes that compete for access rights (keys). Both representations consist of circles, boxes and arrows. It should be kept in mind that what is represented as a circle, a box or an arrow is not an arbitrary matter. Circles designate passive and boxes active system components. Arrows designate relationships between system components.

The circles in these nets are referred to as *channels* and the boxes as *agencies*. (Don't be confused by the use of the term 'channel': there are channels, e.g. in Fig. 71, which also have the label 'channel'. This is not necessarily the case. In Fig. 72, for instance, there is a channel with the label 'keys'.)

Channel-agency net inscriptions are simply descriptions. Inscriptions on channels may describe what objects are in the channels. Similarly, inscriptions on arrows or agencies may indicate when and how agencies work. Channel-agency nets show what components a system is made up of and what relationships exist between which components.

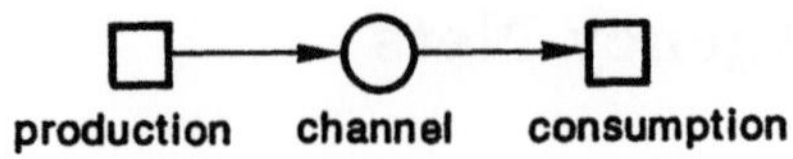

Fig. 71   Rough structure of a producer-consumer system

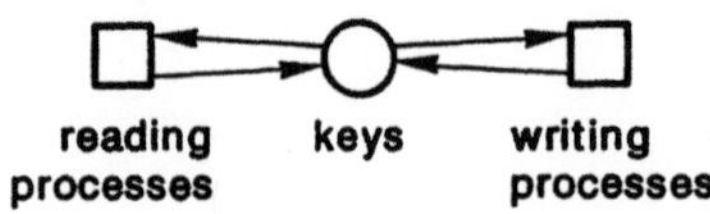

Fig. 72   Rough structure of a system consisting of read and write processes

5.2   Rules

A *channel-agency net* is given on the basis of

- *channels*, represented as circles (O);
- *agencies*, represented as boxes (☐);
- *arrows from channels to agencies* (O⟶☐);
- *arrows from agencies to channels* (☐⟶O);
- (pairs of arrows ☐⟷O can be represented as ☐⟶O) ;
- ordinary language or formal *inscriptions* on channels, agencies and arrows.

For the appropriate use of channel-agency nets the following should be taken into consideration:

In a channel-agency net
- each channel represents a *passive* system component capable of storing, taking on certain states and making objects visible;
- each agency represents an *active* system component that is responsible for the production, transport and change of things;
- arrows indicate a logical connection, physical proximity, access rights and direct links. An arrow never represents a real system component, but rather always an *abstract, logical relationship* between components.

5.3   Further Examples

As a further example let us represent the initial steps in the process of designing a production system such as described in Sect. 2.6. Fig. 73 shows an initial rough structure. The nets in Figs. 74 and 75 are somewhat more detailed.

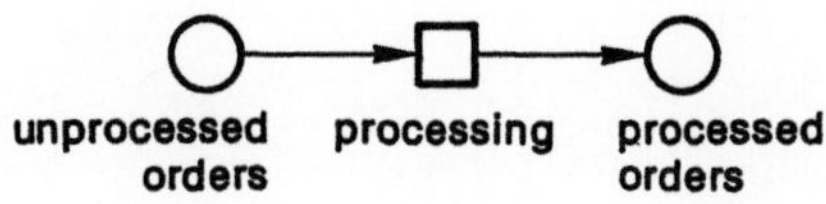

Fig. 73   Rough structure of a simple production system

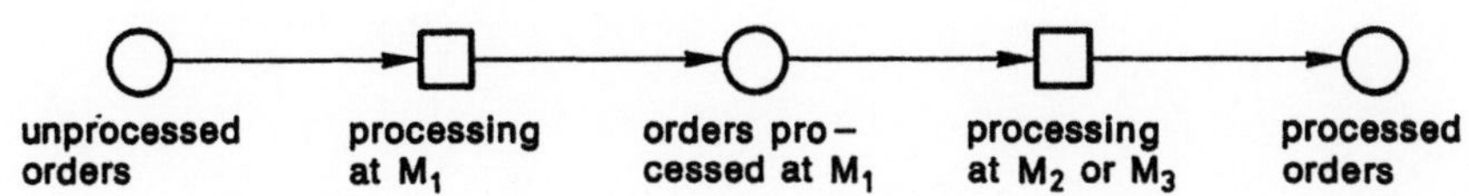

Fig. 74   More detailed representation of the production system in Fig. 73

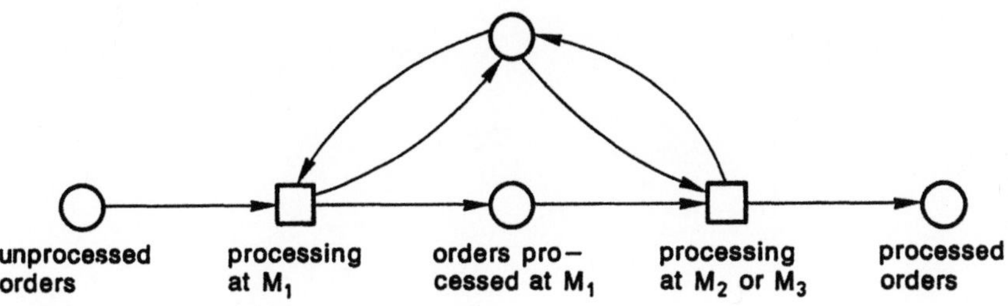

Fig. 75   Addition of a channel to Fig. 74

Problem 28

Design a channel-agency net to model the organization of a service station based on Fig. 38.

Problem 29

Design a channel-agency net modelling the organization of a housing rental market based on Fig. 54.

6 Refinement and Embedding

The nets in Figs.73, 74 and 75 all show the same system from different perspectives, in different sections or in different degrees of refinement. We will see that these nets (and many of the others we have already dealt with) are specifically related to one another, i.e. from a given net another can be systematically derived as a *refinement* or *embedding*.

All of the net models dealt with thus far (condition-event nets, place-transition nets, individual token nets and channel-agency nets) can be refined and embedded. If we speak of nets without wanting to or being able to indicate the corresponding net model, we traditionally use the following descriptive elements, although without any intention of giving special emphasis to places and transitions:

A *net* is given on the basis of
- *places*, represented as circles ($\bigcirc$);
- *transitions*, represented as boxes ($\square$);
- *arrows from places to transitions* ($\bigcirc\!\!-\!\!\!\rightarrow\!\square$);
- *arrows from transitions to places* ($\square\!\!-\!\!\!\rightarrow\!\bigcirc$).

We will deal first of all with refinement and, in particular, discuss what restrictions need to be taken into account in the refinement of nets. After that we will show how nets can be embedded.

6.1 Refinement

We refine a net by substituting an entire net for a place or a transition. The result of this process should, of course, also be a net. A further requirement is that there can only be a connection between the new net component and its net environment if this connection was potentially given in the original net, i.e. a corresponding arrow existed with the same element of the environment in the original net. In this sense, Fig. 74 represents a refinement of Fig. 73.

In graphic net representations a new net component is indicated by a broken-line enclosure.

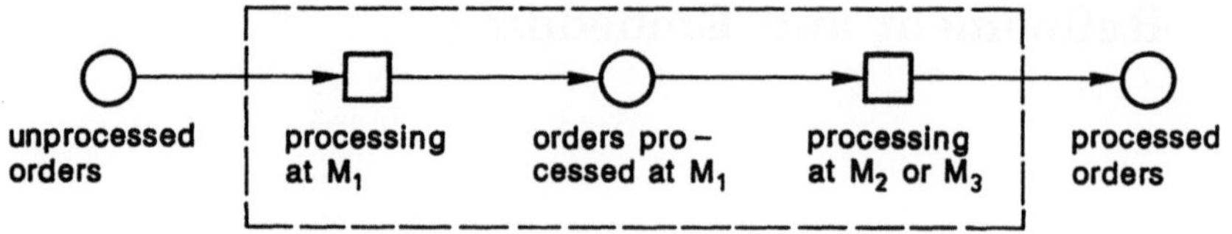

Fig. 76 Graphic representation of a refinement

If we substitute a net for a place, the inserted net appears in an circular enclosure. If we substitute a net for a transition, the inserted net appears in a box-like enclosure (Fig. 76). A refinement has been correctly carried out if the interpretation of the broken-line circles and boxes as places and transitions results in the original net.

In a net A a place p is *refined* by the net B, if B is inserted in the place of p such that for every arrow $x \to y$ from B to A' (or $x \leftarrow y$ from A' to B) it is true that:
-   x is a place of B and y is a transition of A';
-   in A there is an arrow $p \to y$ (or $p \leftarrow y$).
(A' is the net A without the place p.)

In a net A a transition t is *refined* by the net B, if B is substituted for t such that for every arrow $x \to y$ from B to A' (or $x \leftarrow y$ from A' to B) it is true that:
-   x is a transition of B and y is a transition of A';
-   in A there is an arrow $t \to y$ or $(t \leftarrow y)$.
(A' is the net A without the transition t)

A net B is a *refinement* of a net A, if B is a result of the refinement of a number of places and transitions of A. In this case, A is also referred to as a *simplification* of B.

Fig. 77 shows further examples of refinements. Taking into account only the net structure and disregarding markings and inscriptions, the change from Fig. 71 to Fig. 11 is a refinement (but not to Fig. 18!). The same applies to the change from Fig. 72 to Fig. 15 or from Fig.s 73 or 75 (but not from Fig. 74!) to Fig. 24.

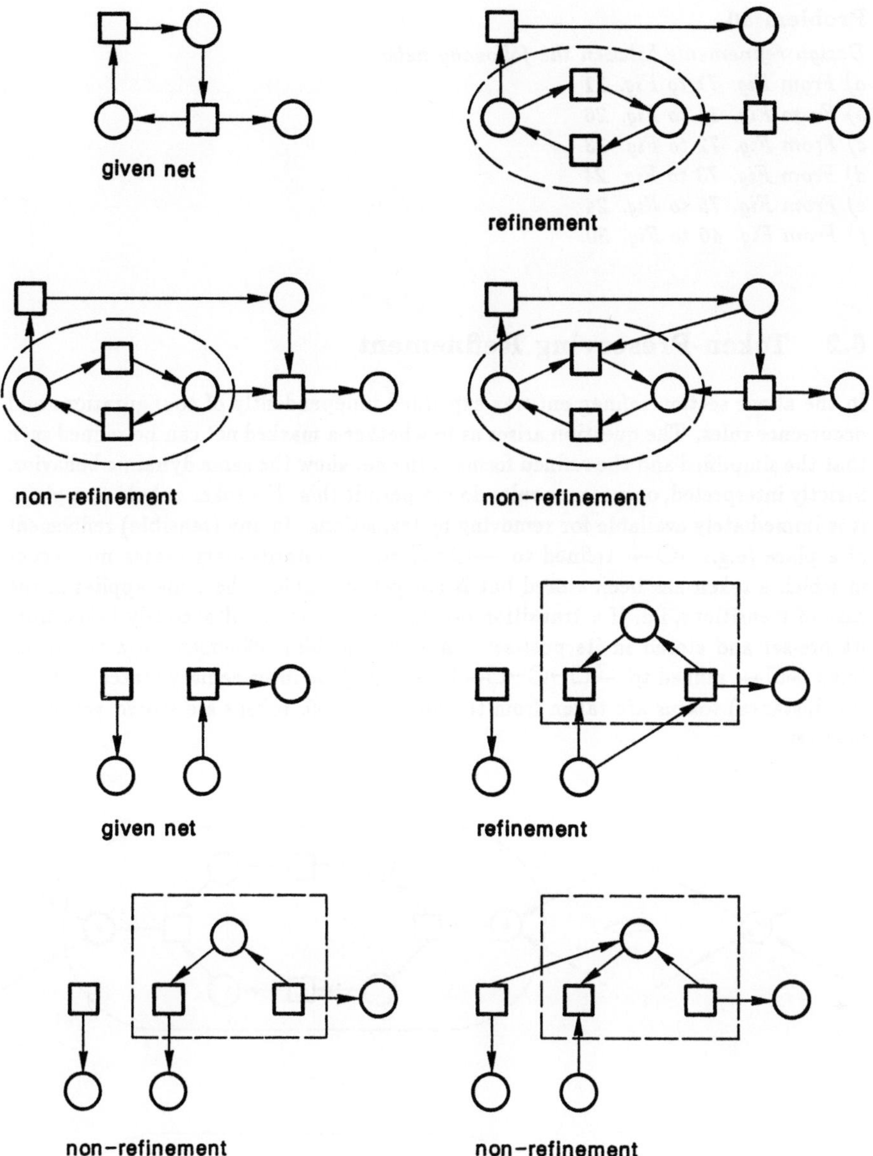

Fig. 77 Examples of refinements and non-refinements

Problem 30

Design refinements between the following nets:
a) From Fig. 71 to Fig. 11
b) From Fig. 11 to Fig. 26
c) From Fig. 72 to Fig. 33
d) From Fig. 73 to Fig. 24
e) From Fig. 75 to Fig. 24
f) From Fig. 40 to Fig. 30.

6.2   Token-Preserving Refinement

In the above section refinement was explained independently of configurations and
occurrence rules. The question arises as to whether a marked net can be refined such
that the simplified and the refined forms of the net show the same dynamic behavior.
Strictly interpreted, occurrence rules do not permit this. If a token is held at a place,
it is immediately available for removing by transitions. In any (sensible) refinement
of a place (e.g.: ⟶○⟶ refined to ⟶○⟶□⟶○⟶) intermediary states may occur
in which a token has been stored but is not yet available. The same applies in the
case of transitions, i.e. if a transition occurs, tokens are simultaneously taken from
its pre-set and stored in its post-set. In any (sensible) refinement of a transition
(e.g.: ⟶□⟶ refined to ⟶□⟶○⟶□⟶) there may be intermediary states in which
(for instance) tokens are taken from the pre-set but no tokens are stored yet in the
post-set.

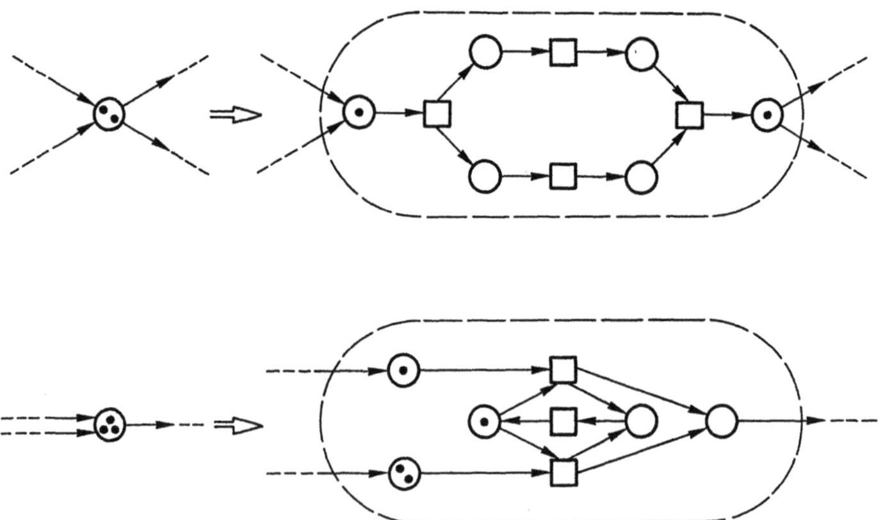

Fig. 78   Token-preserving refinement of places

Although it may not be possible to maintain exactly the same behavior when the net is refined, still, a refined net can behave in a manner 'similar' to the behavior of the original net (the examples referred to as 'sensible' above are of this kind), if certain principles are respected in the process of refinement. A basic prerequisite is that the underlying net be contact-free. Fig. 78 shows a few examples of *token-preserving* refinements.

If, in a place-transition net, a place p is refined by substituting a net component N for it, tokens can be transferred from the rest of the net to N and, from there, tokens can flow into the net from N as a result of the occurrence of "internal" transitions. N has to be able to emit as many tokens as were stored in p in the initial configuration and flowed into N. In contrast to Fig. 78, Fig. 79 shows examples

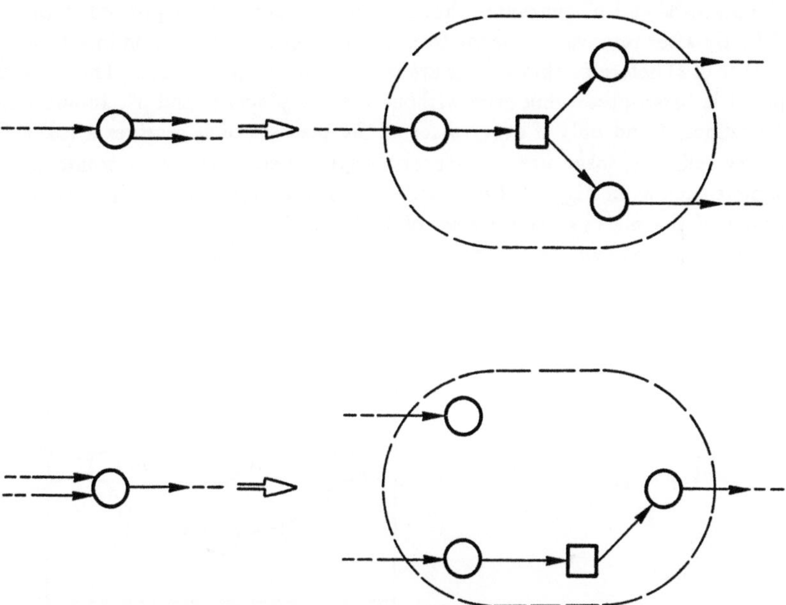

Fig. 79   Non-token-preserving refinement of places

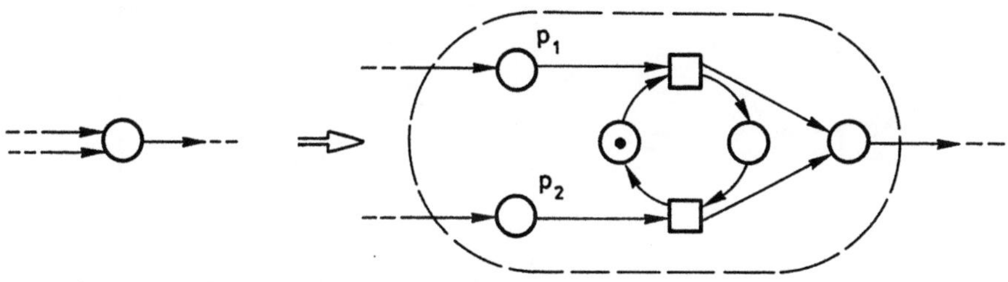

Fig. 80   Refinement of a place that is token-preserving only in special environments

in which this is not the case. Fig. 80 shows a borderline case whose correctness depends on the behavior of the remainder of the net. If p_1 and p_2 contain the same number of tokens (over the long term), the indicated refinement is acceptable. Otherwise, the refined net does not give off enough tokens.

> Let A be a place-transition net. A refinement of a place p of A with a net B is referred to as *token-preserving* if A is contact- free and the initial configuration of p and the tokens flowing into B together correspond to the number of tokens that B is able to emit (on the occurrence of internal transitions).

Fig. 81 shows an example of the token-preserving refinement of a transition.

The places p' and p'' guarantee that a set of tokens will be deposited in the post-set of t only after removal of tokens from the pre-set of t (corresponding to a single occurrence of t) before further tokens are taken from the pre-set of t. The refinement in Fig. 81 is token-preserving even without the two places p' and p''. In addition, t_2 can be refined if and only if every place of the post-set of t_2 receives a token (**Fig. 82**). If we refine t_1, token-preservingness always depends on the environment. In a refinement such as in Fig. 83 it cannot be ruled out that tokens will be taken only from part of the places of the pre-set of t. After that, blockages can occur in the refinement that have no correspondence in the unrefined net.

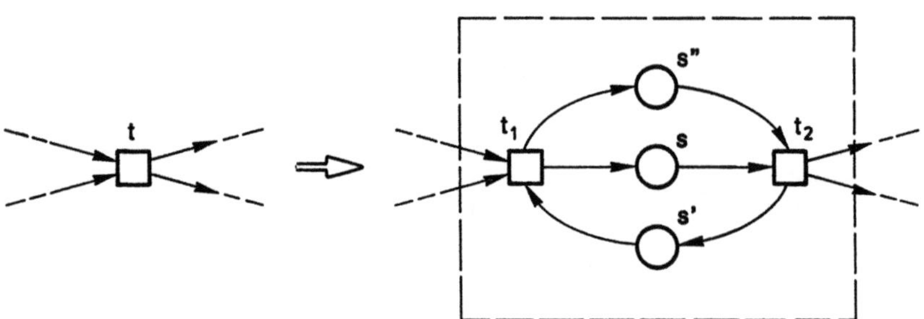

Fig. 81　Token-preserving refinement of a transition

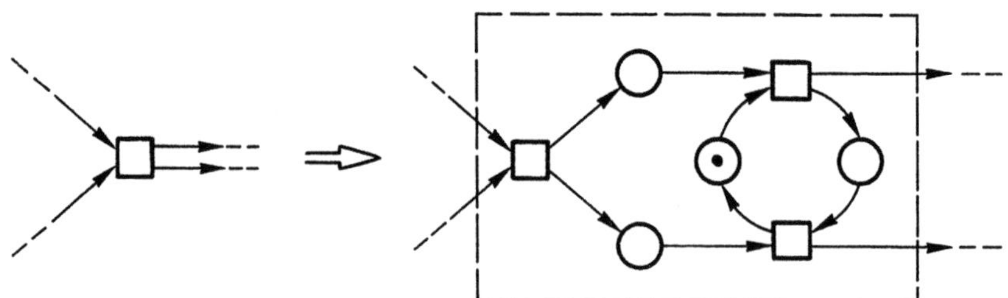

Fig. 82　Token-preserving refinement of a transition

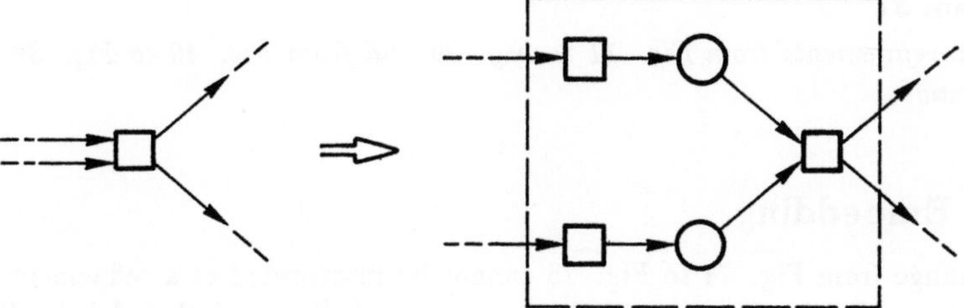

Fig. 83 Refinement of a transition that is token-preserving only in special environments

Let A be a place-transition net. A refinement of a transition t of A with a net B is referred to as *token-preserving*, if A is contact-free and tokens can be taken from all places of the pre- set of t through the occurrence of transitions of B and, after that, all places in the post-set of t receive tokens.

In the case of individual-token nets, a token-preserving refinement must, logically, take into account not just the number (as discussed thus far) but also the identity of tokens. Fig. 84 shows a few examples of token-preserving refinements of individual-token nets.

The criteria for token-preserving refinement are purposely formulated with less precision than in the case of the general rules for net refinement. Whether or not the refinement of a net is token-preserving often depends on the overall net and its use.

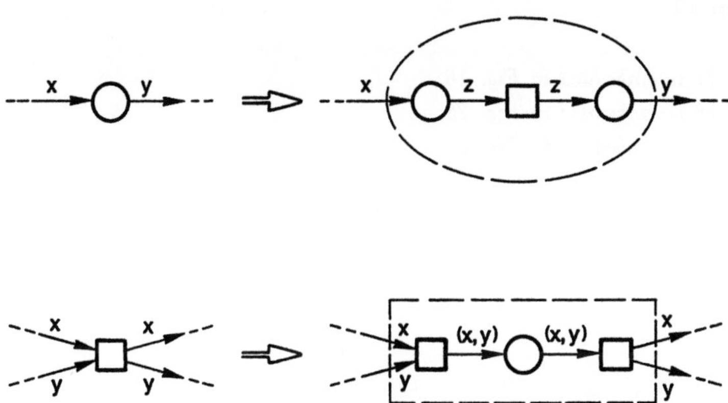

Fig. 84 Token-preserving refinement of individual-token nets

Problem 31

Are the refinements from Fig. 11 to Fig. 26 and from Fig. 40 to Fig. 39 token-preserving?

6.3   Embedding

The change from Fig. 74 to Fig. 75 cannot be interpreted as a refinement. Even if we view only the structure of the two nets and disregard their labels, Fig. 75 contains arrows that have no basis in Fig. 74. If we take into account the channel and agency labels, it inevitably emerges that "operator" is a channel that has been newly incorporated into the net. The net in Fig. 74 is *embedded* in Fig. 75.

While refinement involves replacement of an element of a given net with a new net, in the case of embedding, a new net component is *added* to the given net. The result, of course, should always be a net. A very important example of embedding is the addition of place complements.

A net A is *embedded* in a net B if every place, every transition, every arrow and every inscription of A also occurs in B. In this case, A is also referred to as a *section* of B.

Newly added places, transitions, arrows, tokens and inscriptions can be distinguished, for instance, by using bolder type or different colors.

When a net component is embedded in a given net, the result must of course be a net of the same type (i.e. a condition-event net, a place-transition net, an individual-token net or a channel-agency net). In particular, the tokens of added places must be of the same type.

The information content of a net is increased both in the case of refining and embedding.

Problem 32

Show:

a) *Fig. 11 is embedded in Fig. 18*
b) *Fig. 18 is embedded in Fig. 20*
c) *Fig. 27 is embedded in Fig. 28.*

7 Nets As a Design Method

Various types of nets were introduced in Chaps. 1–4. It was shown on the basis of simple examples how nets can be used to represent the structure and function of different kinds of systems. Obviously, this does not tell us what procedure we need to follow in specific cases in order to succeed in producing an adequate representation of a real or planned system with as little effort as possible. Just as knowing the individual commands of a program language does not enable us to write complicated programs, the 'net language' cannot really be used efficiently without adhering to systematic design procedures.

In this chapter we will show, on the basis of a more complicated and thoroughly realistic example, how nets can be used for the systematic design of computer-based systems. In software engineering terms what we are about to examine is a method for the initial phases of software development (specification phase, requirements engineering) which is particularly suitable for the systematic changeover from specifications imprecisely or incompletely formulated in ordinary language to a precise and implementable model.

7.1 Preliminary Considerations for the Design of Computer-Integrated Systems

Every installed computer is integrated in an environment for which it provides a service (or is supposed to provide a service). The function of a computer in a given situation can only be explained if we take a number of components in its environment into consideration. Thus, if we want to describe what effect a computer has on its environment (and, conversely, what effect the environment has on the computer) we must represent both, i.e. the computer and the environment, or at least those factors that are important for their interaction. A major difficulty in the design of a computer-integrated system lies in the fact that the initial system description is usually formulated in ordinary language and, as such, imprecise and incomplete. From this state it is then gradually formalized, formulated in more detail and completed.

Simplifying things somewhat, the classical situation encountered in systems development involves

- a customer who knows little or nothing about computers (doesn't want to or doesn't need to) and

- a computer specialist who knows very little about his customer's business (only as much as necessary).

In an ideal case, the process of systems development takes place as follows:

- The customer formulates the tasks the intended system is suposed to carry out.
- The computer specialist designs a system, writes manuals and trains personnel.
- The customer tests the system to make sure it works the way it is supposed to (before paying the bill).

As I say, this is the ideal case, and ideal cases are rare. As a rule, the development of systems is a great deal more complicated than this. The primary reason for this is the lack of a means of expression, i.e. a language that

- permits more or less formal descriptions regardless of the type of system involved,
- is readily accessible to the customer,
- makes possible a gradual and systematic transition from ordinary language to formal specifications,
- promotes a problem-oriented structuring of the system and
- makes it possible to formulate and prove system characteristics.

In the following we will present a design method that attempts to match these requirements. It includes all four of the net models we have dealt with. Complex nets are designed from smaller nets derived previously. This is done systematically by means of refinement and embedding.

7.2   An Example

Here is an illustration of how systems can be developed advantageously with the help of Petri Nets on the basis of a somewhat more complicated but realistic example.

Let us take a large store, hereafter referred to only as 'the firm', that sells goods to its customers. A computer-integrated organizational system is to be developed for this company. This presupposes that the processes in the company are precisely described, since this is the only way to clearly assess and formulate what the computer is supposed to do and whether or not the implemented system is working correctly.

Fig. 85 represents the firm as a channel-agency net in an initial rough version. The distribution channel is refined in Fig. 86 and now consists of a channel from the customers to the firm (C–F channel) and a similar channel in the reverse direction. In Fig. 87 we see the concrete configuration of the channels. We can of course ask whether the representation of the telephone as a channel from the customers to the firm is appropriate. The firm can also provide information by telephone to the

Fig. 85   Firm and customers

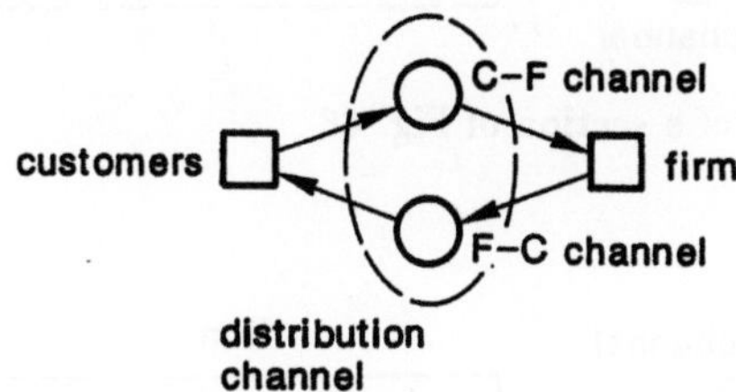

Fig. 86   Refinement of Fig. 85

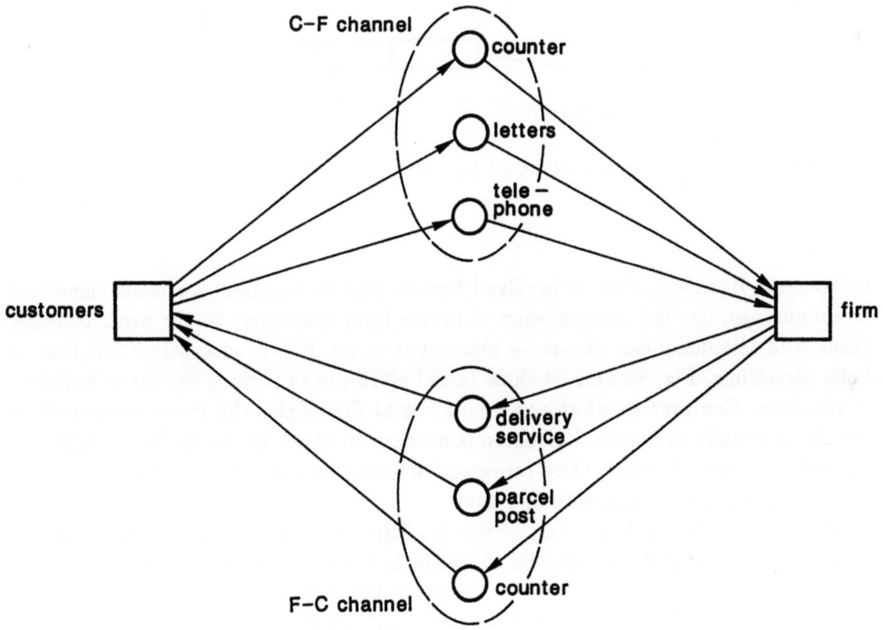

Fig. 87   Refinement of Fig. 86

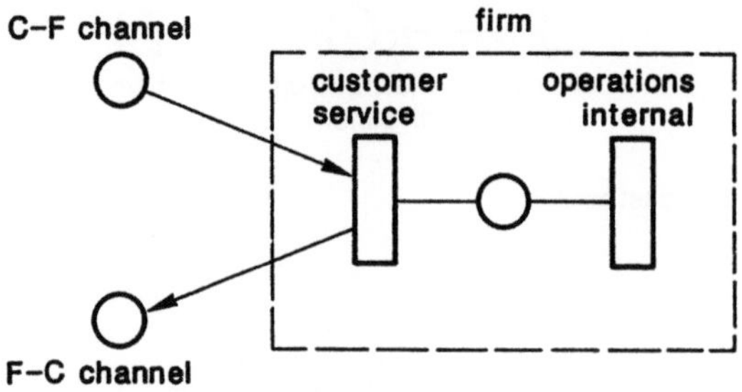

Fig. 88   Refinement of a section of Fig. 86

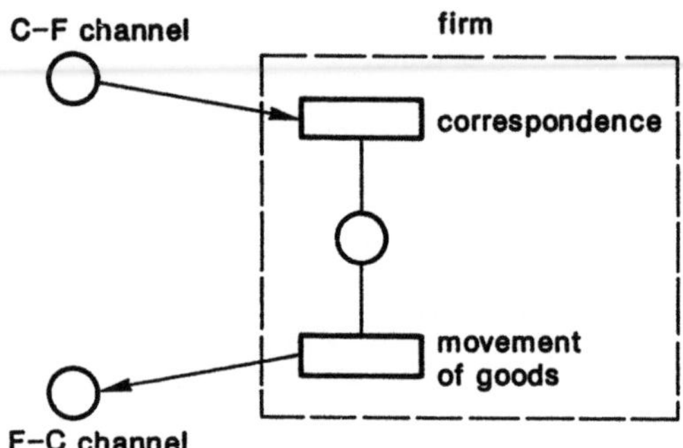

Fig. 89   Refinement of a section of Fig. 86

customers. However, what is involved here is only to represent a specific aspect of telephone use, i.e. the transmission of orders from customers to the firm. In other cases, the telephone can of course also occur as an $F - C$ channel or function in both directions. Fig. 88 and 89 show initial attempts to include the inner workings of the firm. Contrary to what one might fear at first sight, the two representations are not mutually exclusive, i.e. there is no need to think of one as being right and the other wrong. Instead, they represent different aspects of the same system. In Fig. 90 they have a common refinement.

Fig. 91 marks the beginning of the modelling of a new area, i.e. the relations between our firm and the producers of the goods it sells. Aside from what is indicated in Fig. 92, we will not go into the structure of these relations.

Fig. 93 shows what cross-ties between nets established thus far might look like.

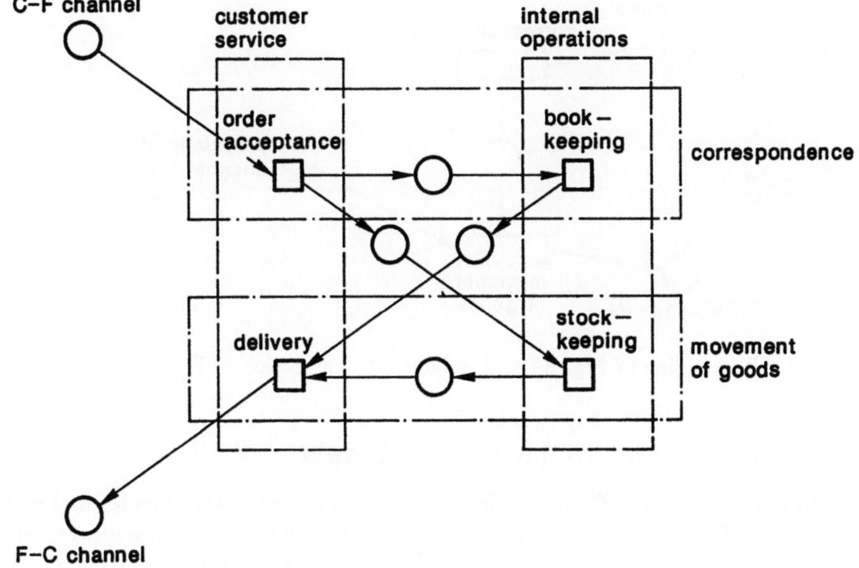

Fig. 90   Joint refinement of Fig. 88 (– – –) and Fig. 89 (– · – · –)

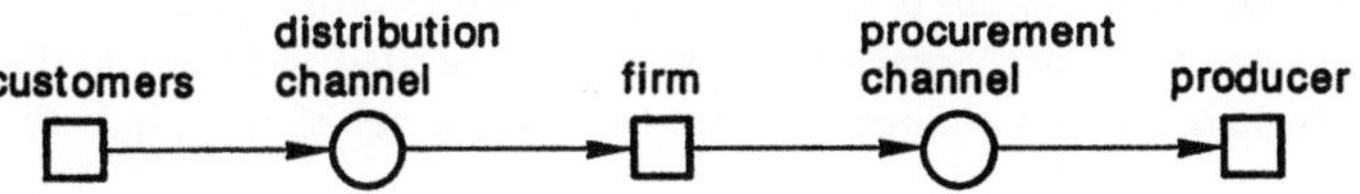

Fig. 91   Embedding of Fig. 85

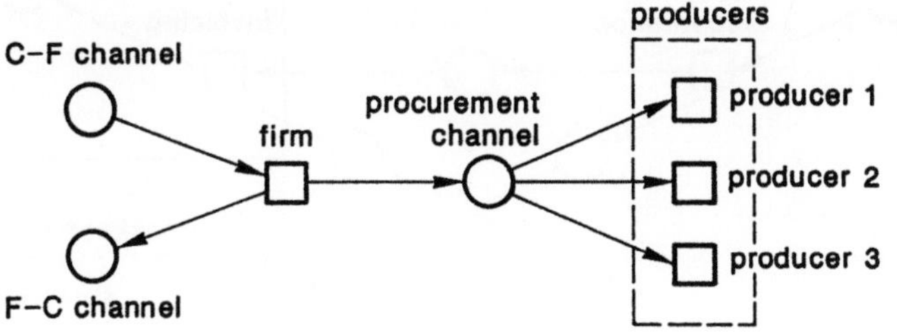

Fig. 92   Refinement of a section of Fig. 91

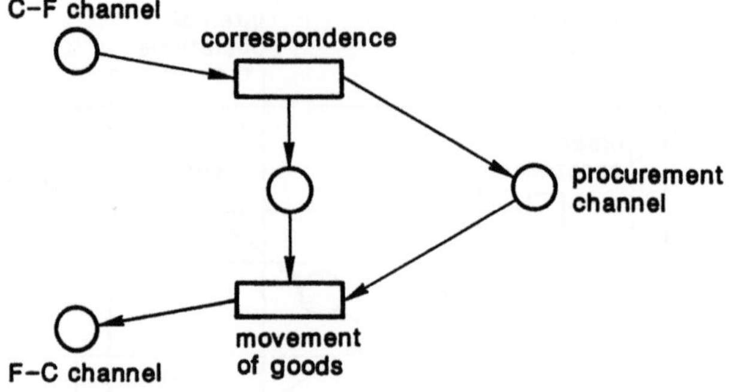

Fig. 93   Embedding of Fig. 89 and refinement of a section of Fig. 92

7.3   The Switch to Other Net Models

As was first indicated in Chap. 4, the example being used in this chapter makes it exceedingly clear that when a large system is being modeled, channel-agency nets have to be used initially, from which more detailed nets are gradually derived on the basis of refinement and embedding. At some point a stage may be reached at which the mere further breakdown of the overall system into sub-systems is no longer of interest and that we will want to provide a more precise model of dynamic behavior.

In the context of our example there have been no nets thus far in which the dynamic behavior of the system is described meaningfully (appropriate to the problem in question) using the occurrence rules we are acquainted with. Using Fig. 90, we only need to take another refinement step to attain a net of this kind. (Fig. 94)

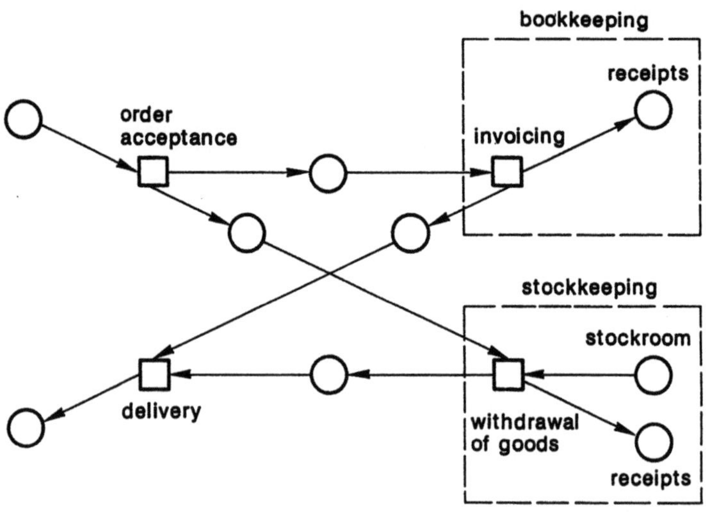

Fig. 94   Refinement of Fig. 90

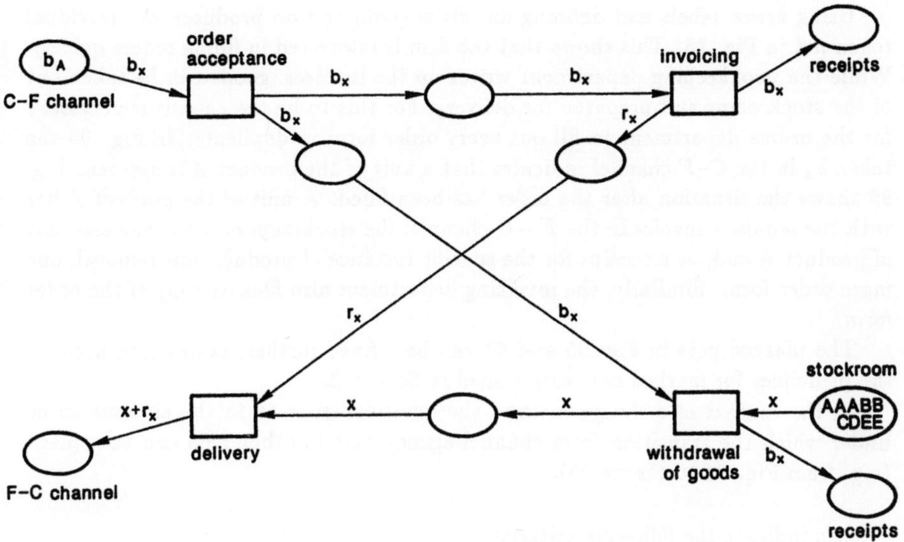

Fig. 95   Refinement of Fig. to an individual-token net

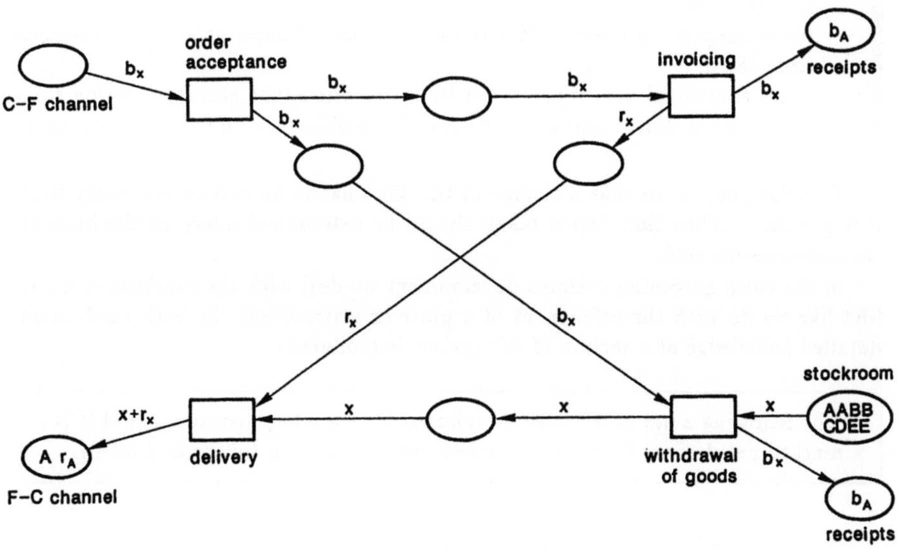

Fig. 96   Situation following completion of the order indicated in Fig. 85

Using arrow labels and defining an initial configuration produces an individual token net in Fig. 95. This shows that the firm is interested in filling orders quickly. While the bookkeeping department writes up the invoices, goods can be taken out of the stockrooms and prepared for delivery. For this to be the case, it is necessary for the orders department to fill out every order form in duplicate. In Fig. 95 the token b_A in the C–F channel indicates that a unit of the product A is ordered. Fig. 96 shows the situation after the order has been filled. A unit of the product A lies with the requisite invoice in the $F - C$ channel, the stockroom now has one less unit of product A and, as a receipt for the specific instance of product-unit removal, one more order form. Similarly, the invoicing department also files its copy of the order form.

The marked nets in Figs.95 and 96 can be refined further, taking into account the principles for marked nets established in Sect. 6.2.

In the context of a design method the question arises as to the circumstances under which the transition from channel-agency nets to other nets can take place (e.g. from Fig. 94 to Figure 95).

We can indicate the following criteria:

The switch from a channel-agency net to a condition-event, place- transition or individual-token net can be made if, at the chosen abstraction level,
-   every channel represents a functional unit that stores objects but does not change them;
-   every agency represents a functional unit that changes objects but does not store them;
-   every functional unit represented by a transition that goes into action takes objects from all its input channels and places objects in all its output channels.

This does not mean that a change of this kind has to be carried out every time it is possible. When this step is taken should be determined solely on the basis of the problem involved.

In the rules governing systems development we deal with the marking of a net just like we do with the refinement of a place or a transition. In both cases more detailed knowledge of a section of the system is produced.

In the following a net B will also be referred to as a refinement of a net A if B is a marked net derived from an unmarked net A by adding tokens and labels.

7.4   Additions to the Example

Let us now examine a few additions to our example. The purpose of these additions is to take into account individual customers, to define order acceptance and product delivery procedures, as well as to check customer creditworthiness.

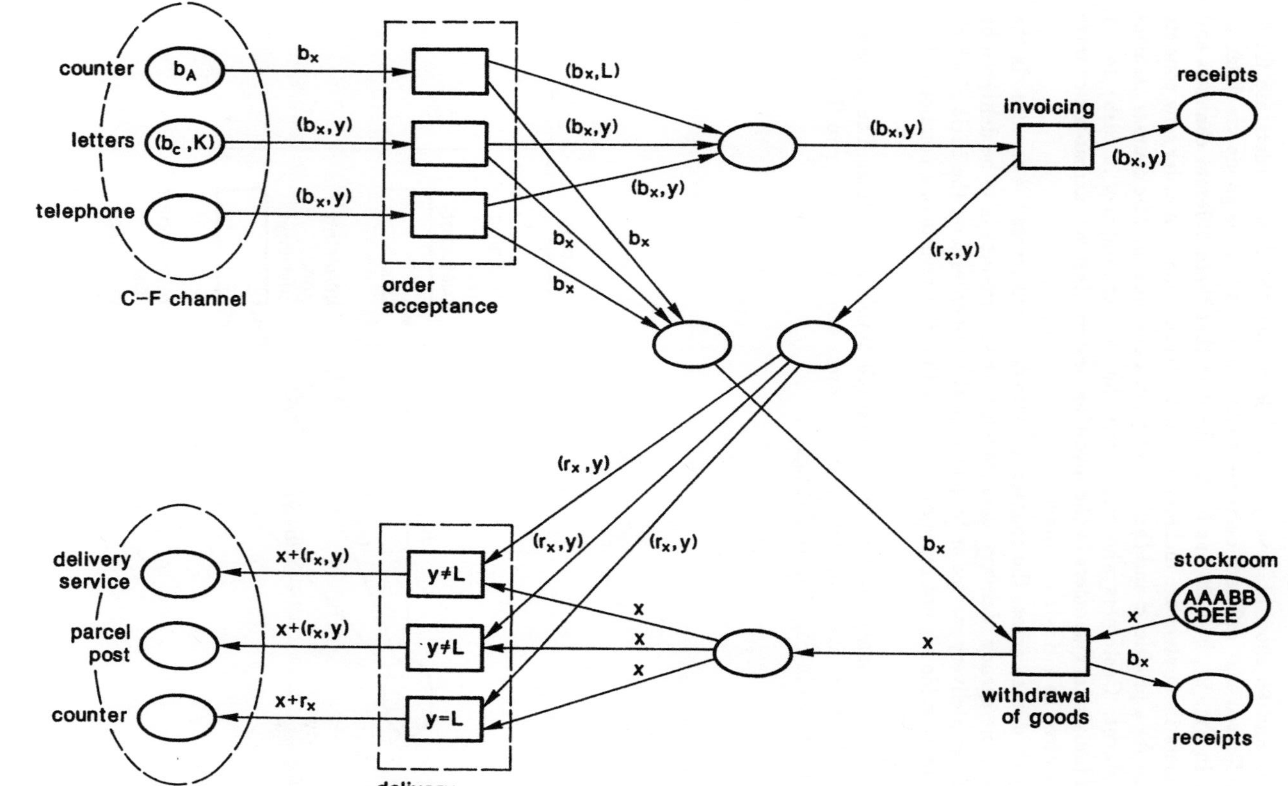

Fig. 97   Joint refinement of Fig. 85 and a section of Fig. 77

Fig. 95 and 96 show at a very general level how the firm's orders are filled. Fig. 87 indicated that different customers can order and receive goods via different channels. In Fig. 97, Fig. 95 has been refined so that these different channels and the customers themselves are included in the representation. An order now consists of a request for a product x and the indication of a customer y. The counter assumes a special status. Customers who buy across the counter (and pay in cash) do not appear as individual customers in the invoicing process. Instead, across-the-counter sales are viewed as a single customer.

Products ordered across the counter are also delivered across the counter. In the other cases, the agency 'delivery' decides whether the product is to be delivered by the firm's own delivery service or by parcel post. The criteria on the basis of which this decision is made are not represented. Note that this results in a conflict in Fig. 97.

A further (and final) addition to our example involves invoicing. The purpose, here, is to check the creditworthiness of customers and not to deliver to 'bad' customers. As shown in Fig. 98, we distinguish

1) new,
2) good and
3) "bad" customers.

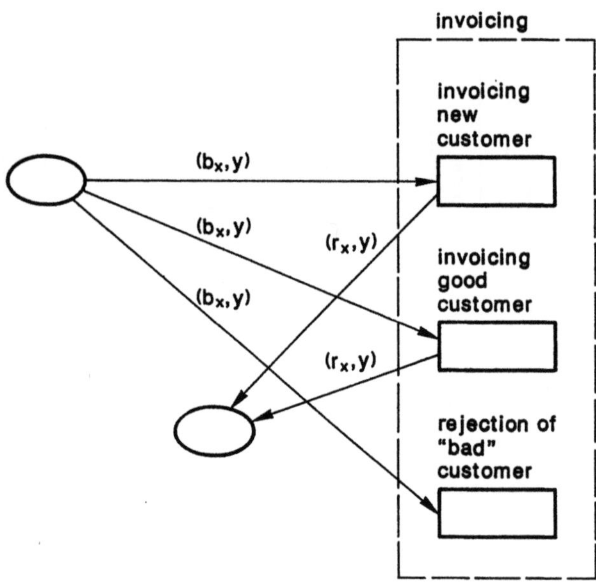

Fig. 98   Refinement of a section of Fig. 97

82

Fig. 99 gives a detailed account of how customers are served. The central element is a list including every customer the firm has had thus far (double entries are ruled out). Customer status is indicated after the name of each customer. The status is either positive (+) or negative (−) and indicates whether the customer is creditworthy or not. New customers are initially seen as being creditworthy. Across-the-counter customers are dealt with here like the other creditworthy customers.

The assumption is made that the number of uncreditworthy customers will be small. The advantages of rapid customer service discussed in Sect. 7.3 outweigh the expense incurred by having to put goods ready for delivery back into storage when a 'bad' customer is rejected.

This section dealt with the refinement of marked nets. As can easily be seen, these refinements are token-preserving.

The customer list in Fig. 91 is a component that can be seen as part of an implementation specification.

Problem 33
What configuration occurs when the two orders indicated in Fig. 97 are filled?

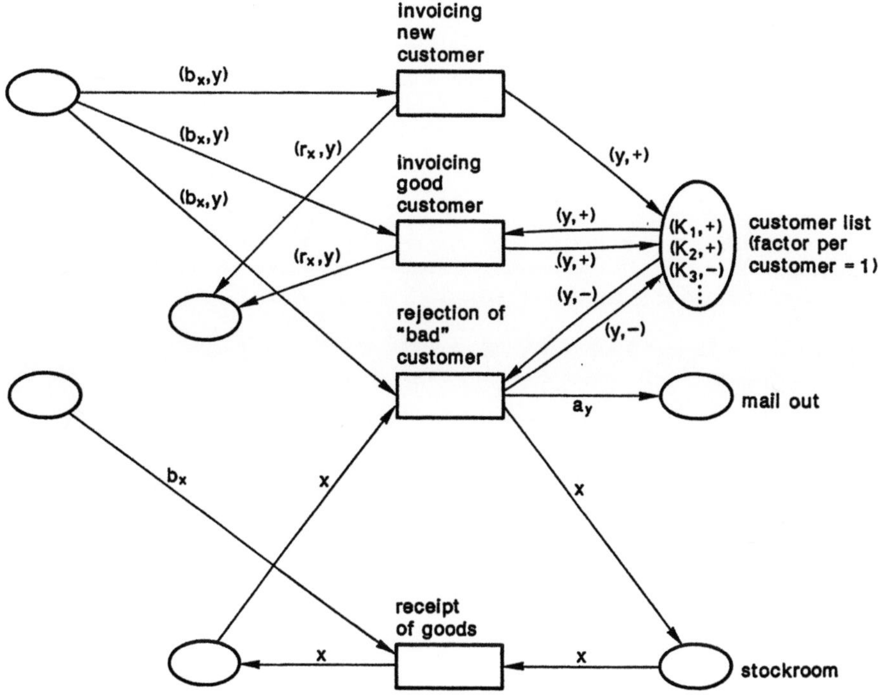

Fig. 99   Embedding of Fig. 91

Solutions

Problem 1

The channel is replaced by two storage cells in sequence. The channel itself is not represented in Figure 11, but rather the condition that says whether or not the channel is empty. Accordingly, we stipulate conditions indicating whether or not the storage cells are empty. The fact that the storage cells are arranged sequentially means that objects can be transferred from the first storage cell to the second. Thus, in the model there must be an event, which, when it occurs, causes the condition 'storage cell 1 occupied' to be unfulfilled and the condition 'storage cell 2 occupied' to be fulfilled.

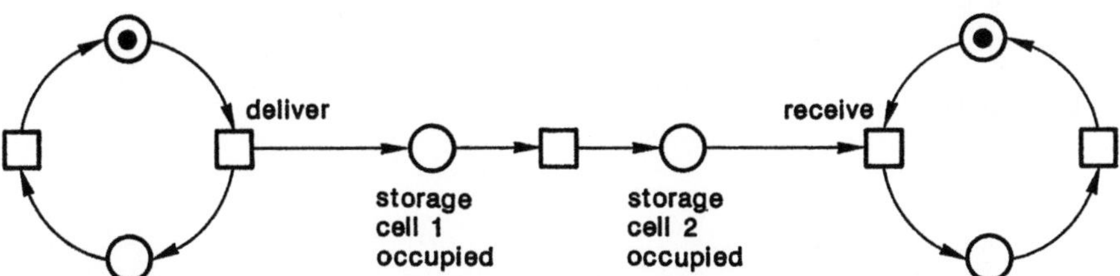

Figure 100

Problem 2

a) The condition 'it is fall or winter' is symmetrical with the condition 'it is winter or spring' indicated in Figure 14. When fall begins 'it is fall or winter' becomes true. When spring begins (identical with the end of winter) it becomes false.

b) When fall begins 'it is not summer' becomes true and remains true until the beginning of summer.

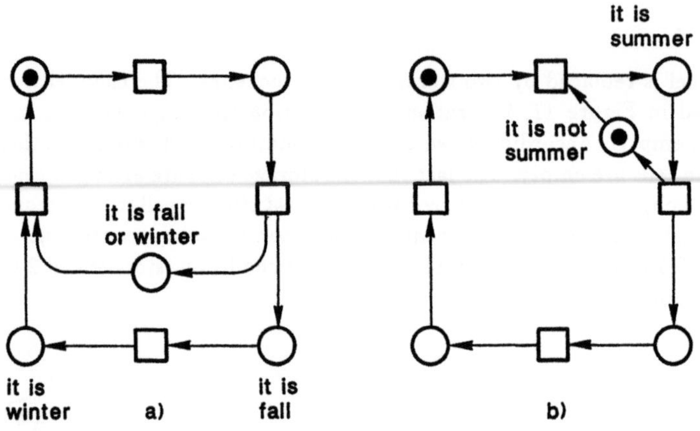

Figure 101

Problem 3

a) 'it is summer or winter' is true twice and false twice in the course of a year. Note that 'it is summer or winter' is the common precondition for the 'beginning of fall' and the 'beginning of spring'. There is never a conflict, since the indicated events are never activated at the same time.

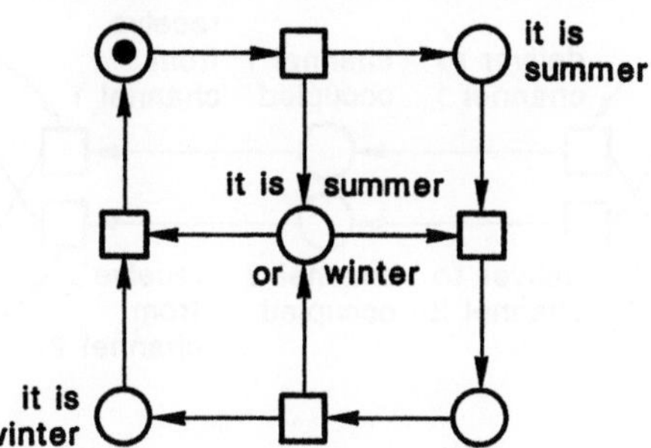

Figure 102

b) Design a third cycle including the condition 'key is available'.

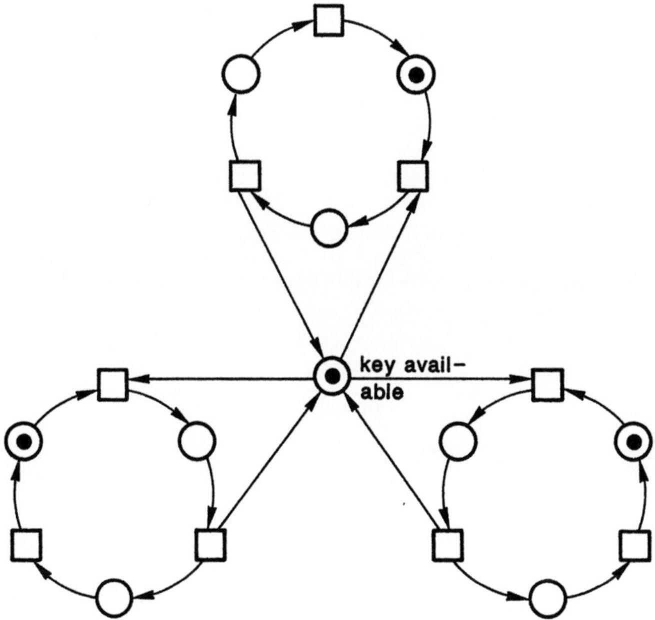

Figure 103

Problem 4

When the system is ready for sending it must be decided what channel is to be occupied. If both are unoccupied there is a conflict. The same applies when the system is ready for removal.

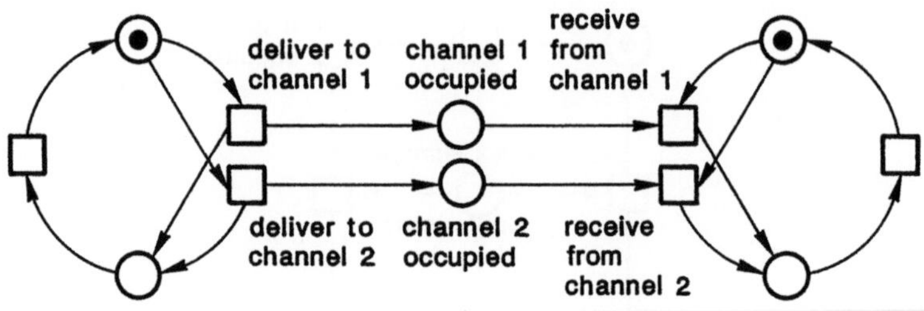

Figure 104

Problem 5

Figures 14 and 15 are both contact-free.

Problem 6

The complements of the four 'external' conditions 'it is spring', 'it is summer', etc. are 'it is not spring', 'it is not summer', etc. The complement of 'it is winter or spring' is 'it is neither winter nor spring' or 'it is summer or fall'.

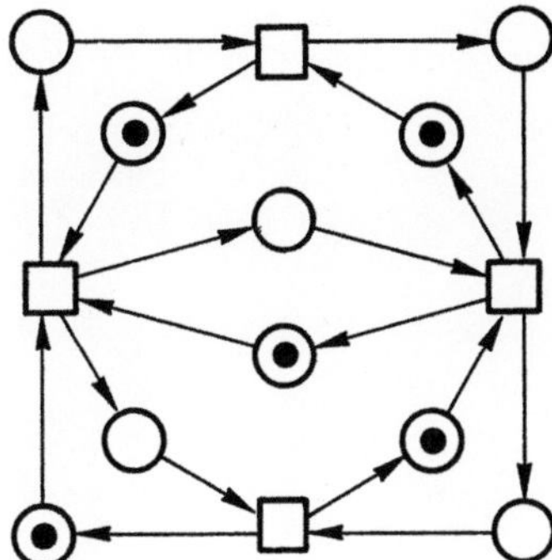

Figure 105

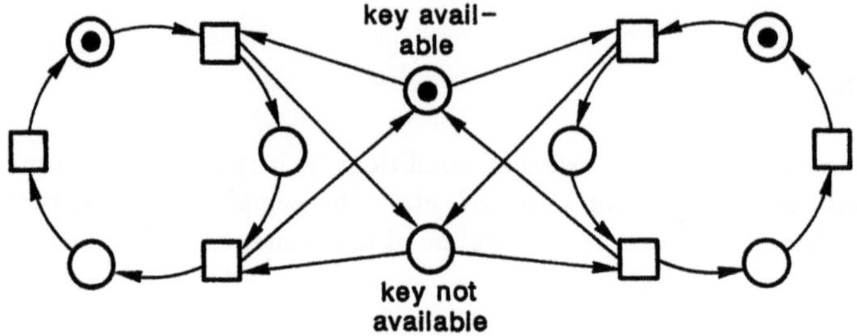

Figure 106

Problem 8

We will use the following abbreviations for the graphical representation of the processes:

Figure 14: Conditions *1* and *5* are fulfilled in the case indicated. Both are prerequisites for the occurrence of *a*. Let us design a process in which *a* occurs twice
and every other event once.

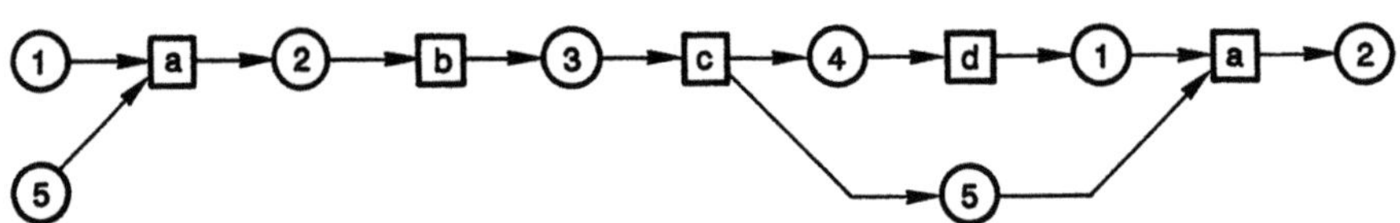

Figure 107

90

Processes corresponding to the additions asked for in Problems 2 and 3a have the following structures (note that for Figure 14 there is exactly one unlimited (non-ending) process and that every other process is an initial element of it; the same applies for all systems in Problems 2 and 3a):

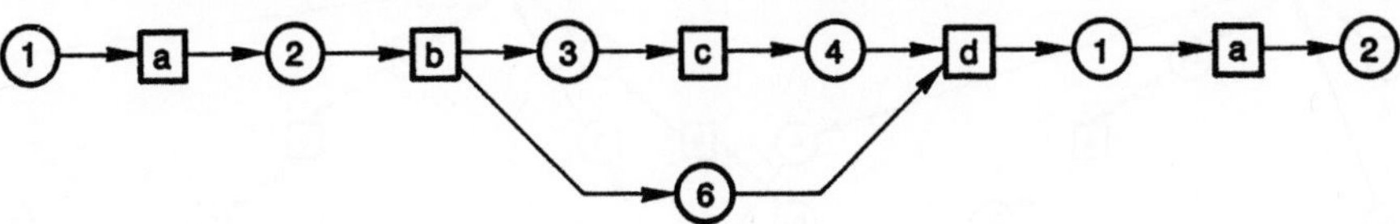

Problem 2a

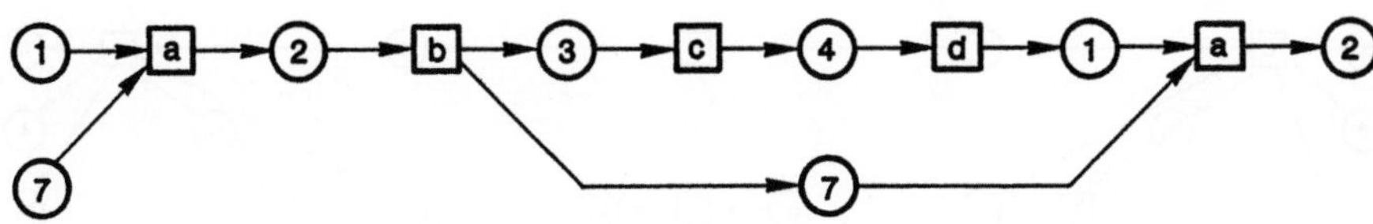

Problem 2b

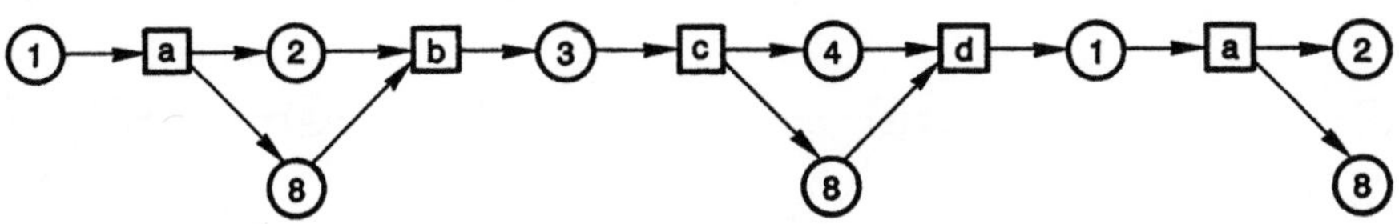

Problem 3a

1 = it is spring
2 = it is summer
3 = it is fall
4 = it is winter

a = beginning of summer
b = beginning of fall

5 = it is winter or spring
6 = it is fall or winter
7 = it is not summer
8 = it is summer or winter

c = beginning of winter
d = beginning of spring

Figure 108

Problem 9

Three examples of processes relating to Figure 20:

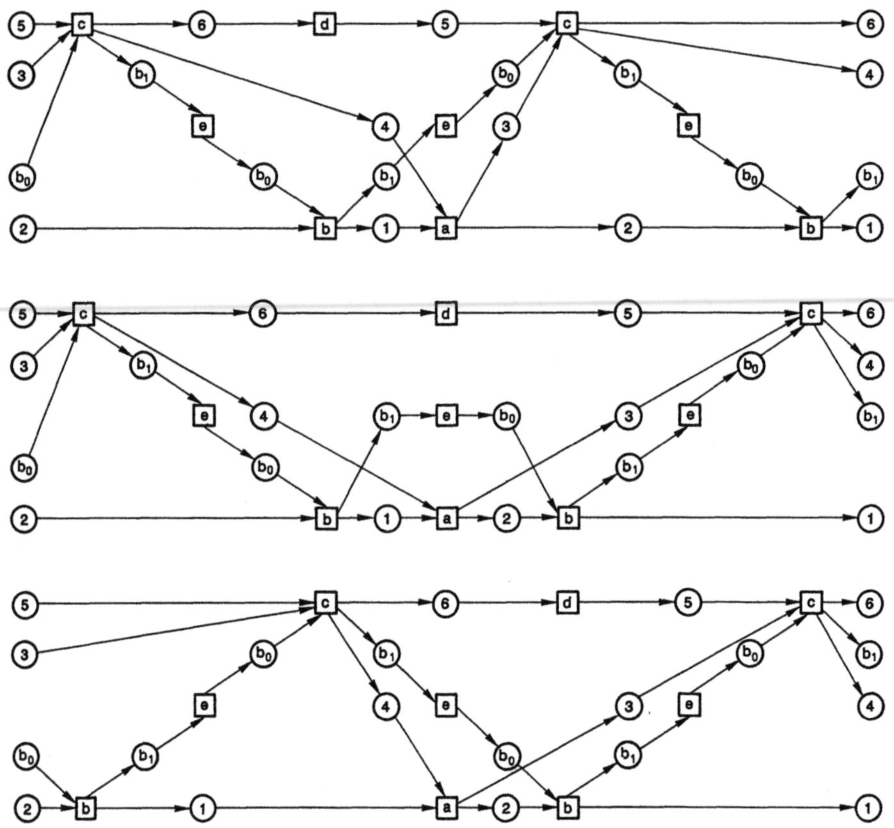

Figure 109

Similar to Figure 14, Figure 18 has one and only one unlimited non- ending process and every other process is an initial element of it (e.g. the process indicated in Figure 21). The same is not true in the case of Figure 11, since here a conflict can occur (between 'produce' and 'receive'). As such, the three processes indicated above are not initial elements of a single long process. The three processes are similar to Figure 21 with regard to their upper and lower lines, where the events c and d or d and a are repeated alternately. The synchronization between these lines is different in each case.

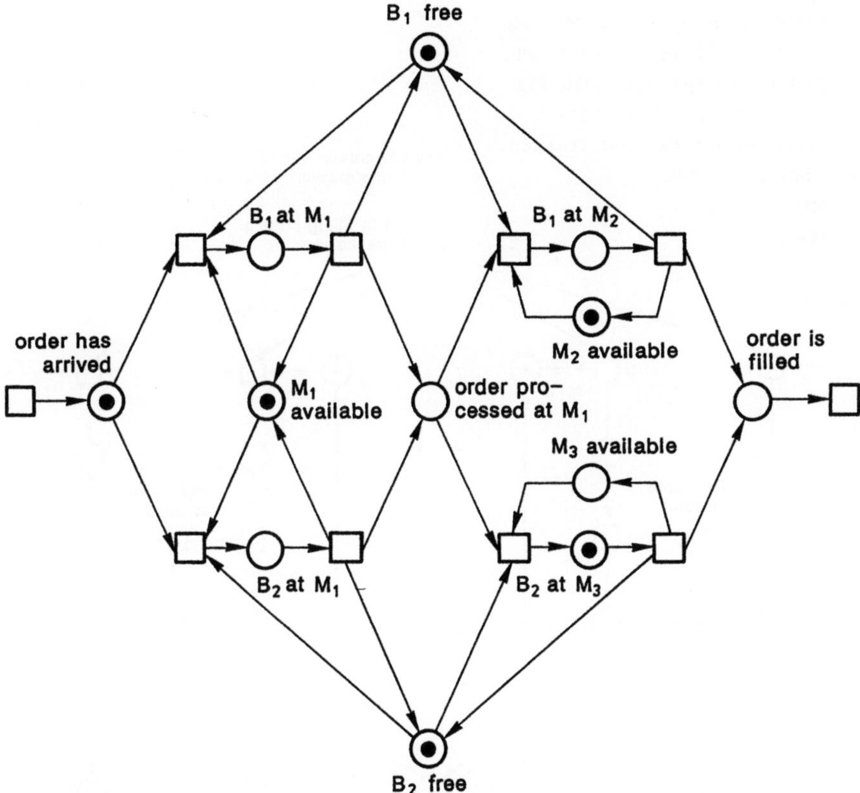

Figure 110

The condition 'M_2 free' is a complement of 'O_1 to M_2'. In Sect. 1.4 it was shown that the introduction of a complement does not influence system behavior. The same applies to 'M_3 free' as a complement of 'O_2 to M_3'. 'M_1 free' is not the complement of any condition from Figure 24. Its absence would permit O_1 and O_2 overlapping access to M_1.

Problem 11

a) There is a total of four parking
   spaces. The entire course of
   events is mapped out for each
   of them. There is a new conflict by comparison with Figure 25, given the fact that now
   every pump can be reached
   from more than
   one parking
   space.

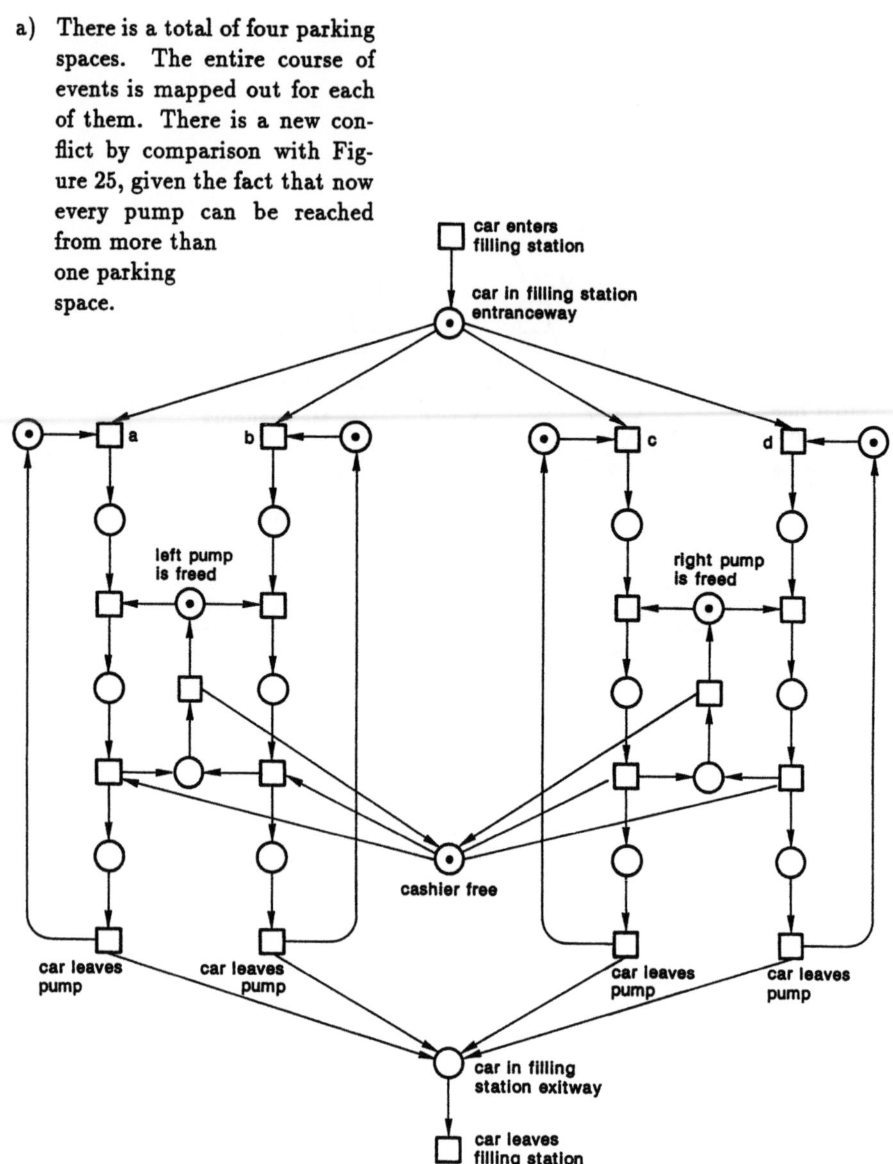

a Car drives up to left of left pump
b Car drives up to right of left pump
c Car drives up to left of right pump
d Car drives up to right of right pump

Figure 111

b) A cycle of events is mapped
out for each of the two cashiers
in which they are paid and
then the pumps are freed for
the next customer. New conflicts arise (between a and b)
in that in paying a choice has
to be made between the two
cashiers.

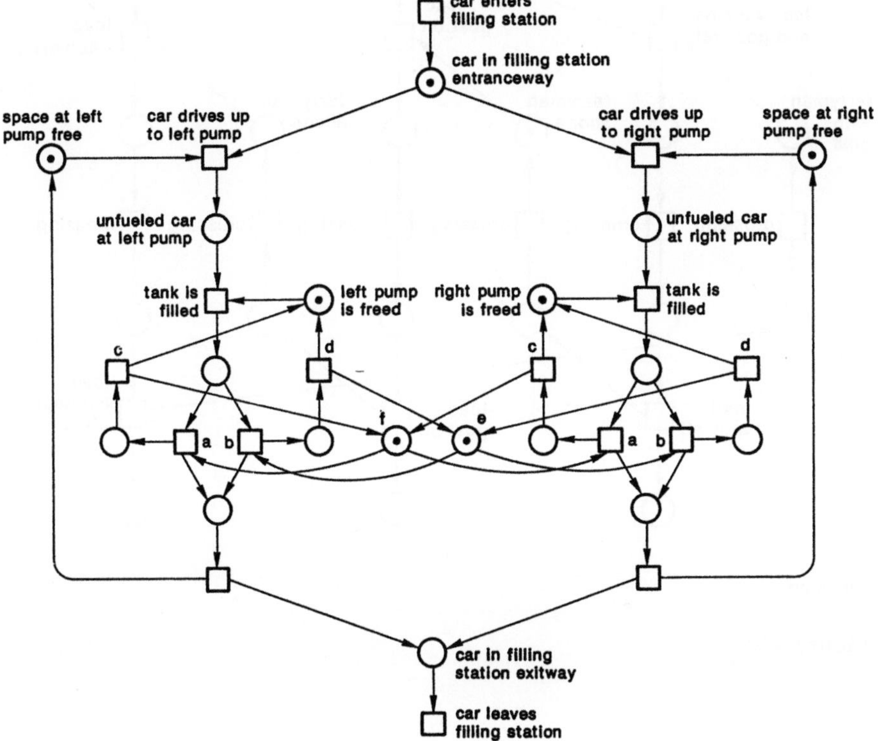

a Customer pays cashier I
b Customer pays cashier II
c Cashier I frees pump
d Cashier II frees pump
e Cashier I free
f Cashier II free

Figure 112

Problem 12

The problem is solved by taking the goat back across the river at one point.

other side

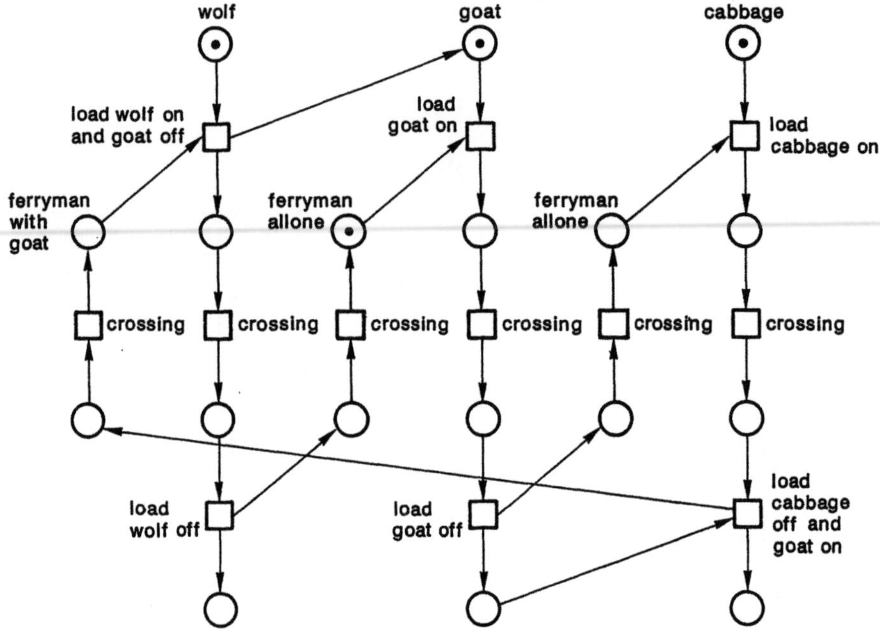

this side

Figure 113

Problem 13

The sending and receiving of several objects at the same time can be easily represented by using weighted arrows. The restriction of recipient consumers to just one is modelled by using a capacity limit.

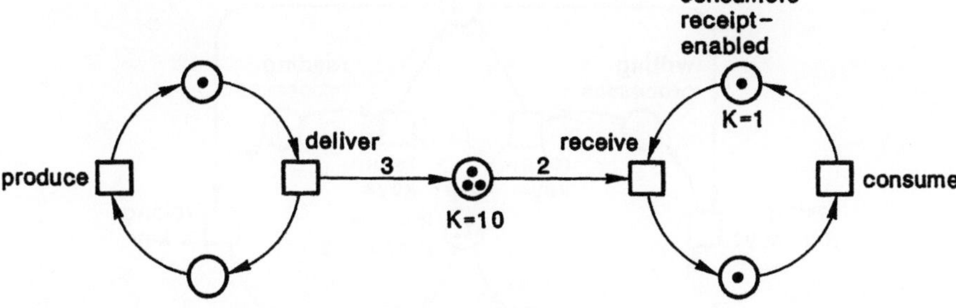

Figure 114

Problem 14

We now have four processes with equal priorities. Since they can all read at the same time, four keys are required.

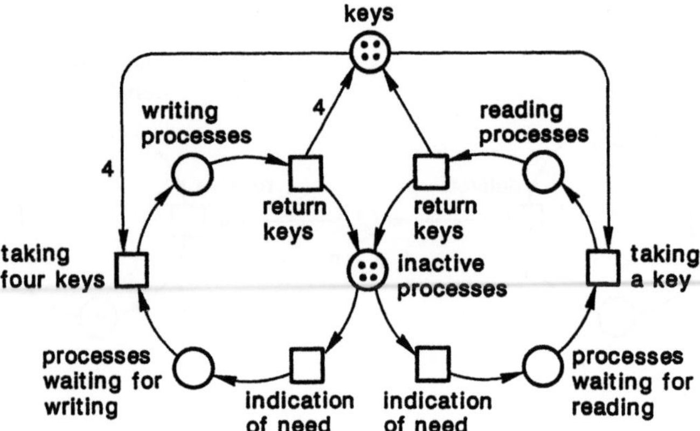

Figure 115

Problem 15

a) To design complements it is necessary to use capacities. Only the capacity of the channel is known (10). Since there is never more than one token at a time in the producer cycle in Figure 114, a capacity of *one* is sufficient. As many as two tokens may be present per place. One of the places has a capacity limit of one. Thus, its complement also has a capacity of one.

b) Only the complements of the channel and the place p are needed to make this net contact-free. The complements of the three other places are not necessary.

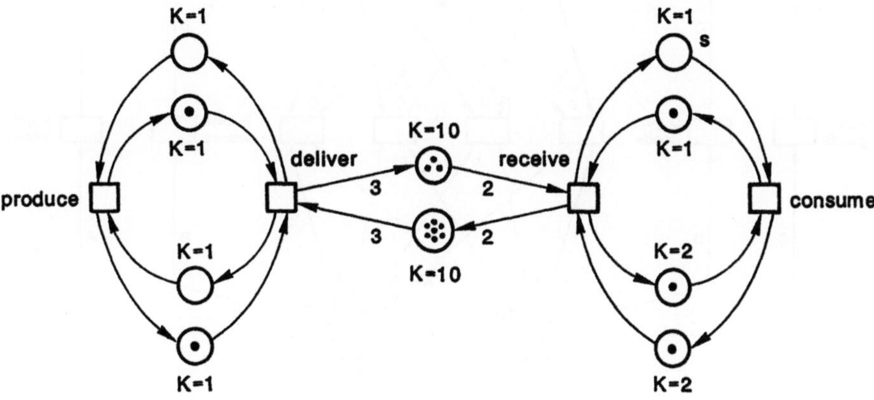

Figure 116

Problem 16

a) Eight different configurations are possible in Figure 41.

b) In the case of two dealers and two customers only four different configurations
    are possible.

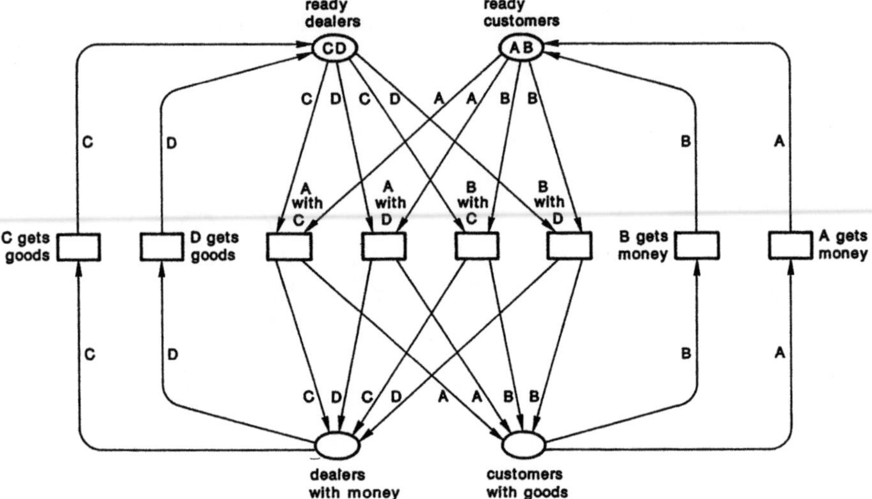

Figure 117

Problem 17

A new place is introduced that has a token when it is winter or spring. This place
thus behaves likes the condition 'it is winter or spring'. It receives a token at the
beginning of winter and loses the token at the end of spring. The token is indicated
by *wos* (= "winter or spring").

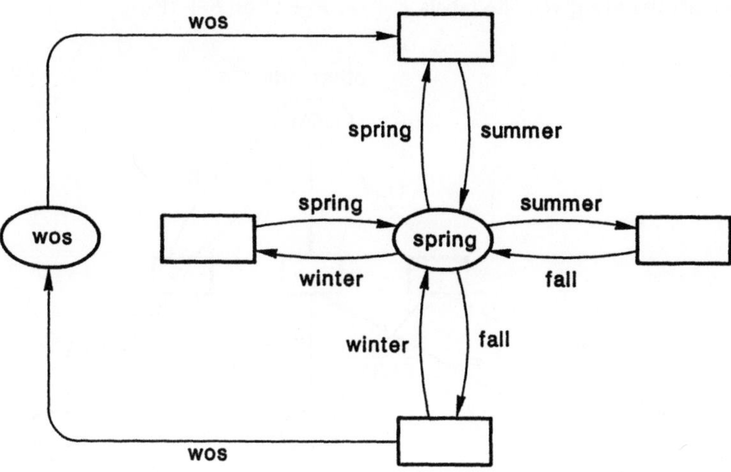

Figure 118

Problem 18

We will use the following abbreviations for the different tokens:
- G for goat
- C for cabbage and
- W for wolf.

The ferryman F is accompanied in some cases by a second object (the goat, the cabbage or the wolf). Since this situation is modelled with two tokens, the corresponding arrows are given the labels $F + G$, $F + C$ or $F + W$.

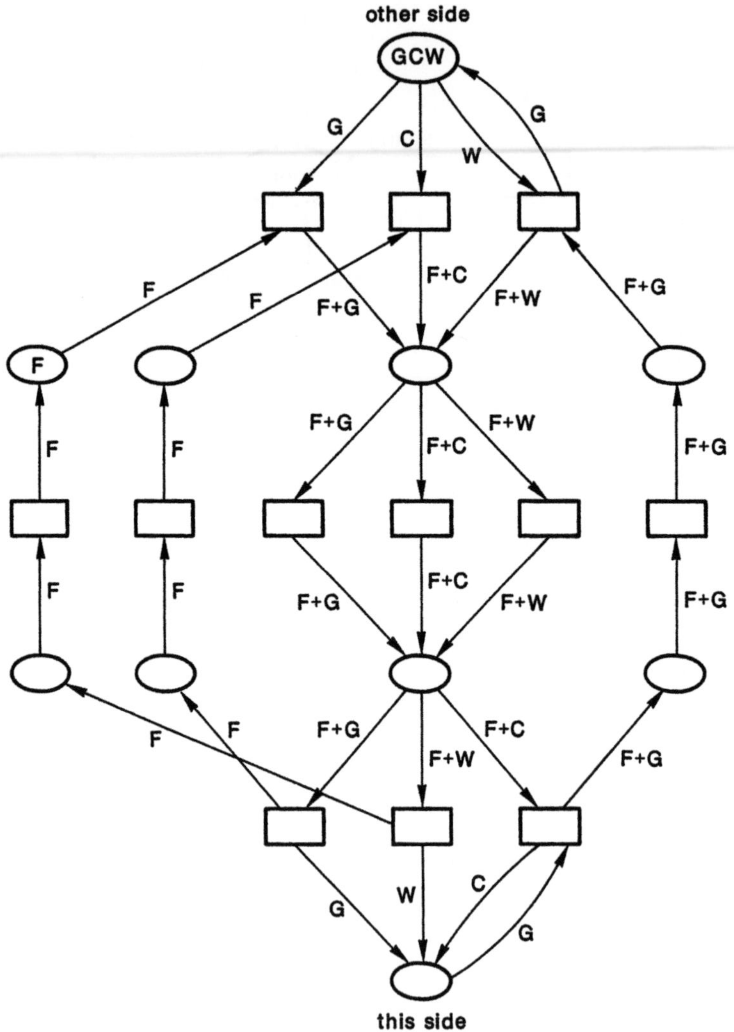

Figure 119

Problem 19

We will use the abbreviations introduced in Figure 25, i.e. B_1, B_2, M_1, M_2, M_3 as well as O (for 'order') as individual tokens. To indicate what employee carried out a specific order, we will add the token B_1 or B_2 to the token O. Thus, we may have tokens such as (B_1, O) or (B_2, O).

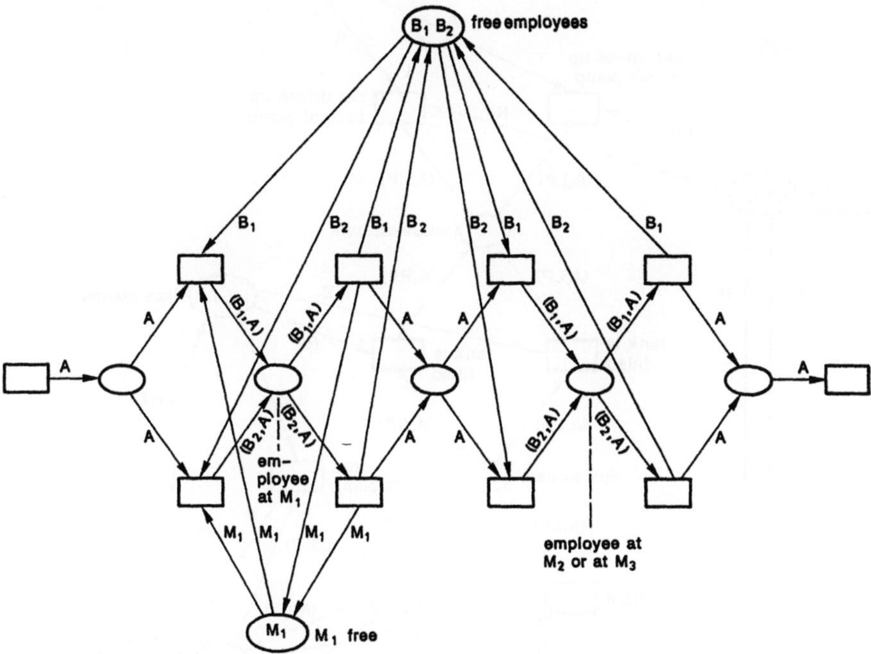

Figure 120

Problem 20

As was the case in Problem 19, it is helpful here to use pairs of objects as individual
tokens.

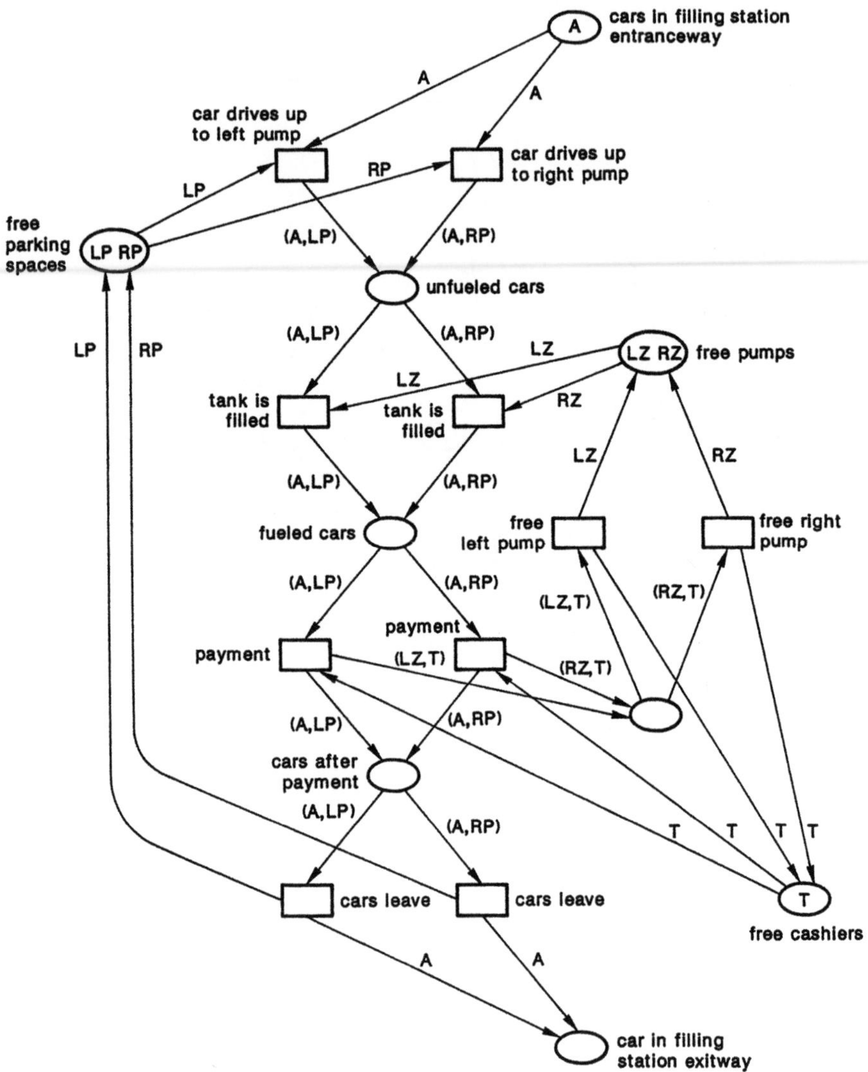

Figure 121

Problem 21

Equality and inequality of variables is only important when the arrows in question begin or end at the same transition. In order to use as many variables as possible, we can employ specific variables for every transition, i.e. variables that only occur in connection with transition in question. In order to use as few variables as possible, on the other hand, we can employ the same variables on the arrows of the various transitions.

a) Six different variables can be used.

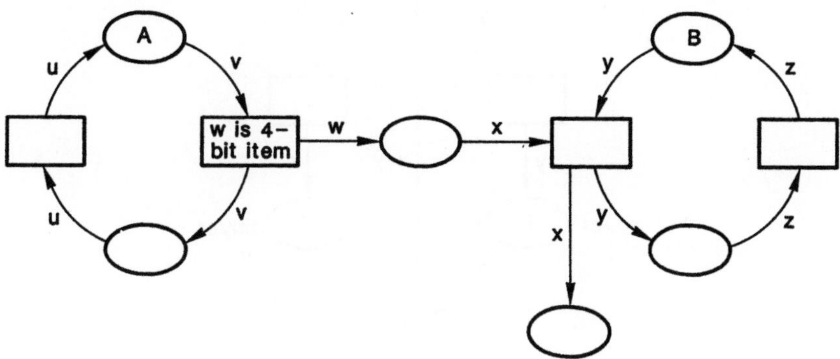

Figure 122

b) Two different variables are enough.

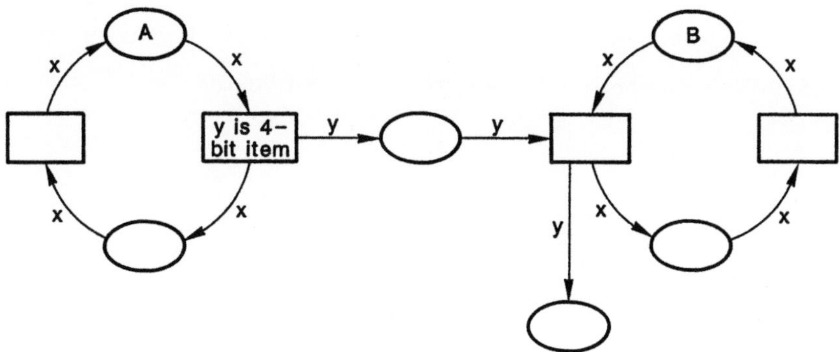

Figure 123

Problem 22

We will choose a configuration in which a single person acts as both a buyer and a seller and all others are inactive. When two partners have made a deal they both go into a non-active state. This is indicated by means of the arrow label '$x + y$'.

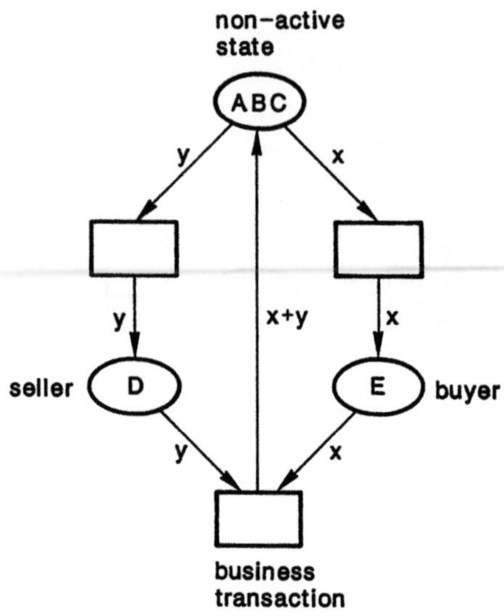

Figure 124

a) As in the solution to Problem 19, we will use pairs of operators and orders as tokens.

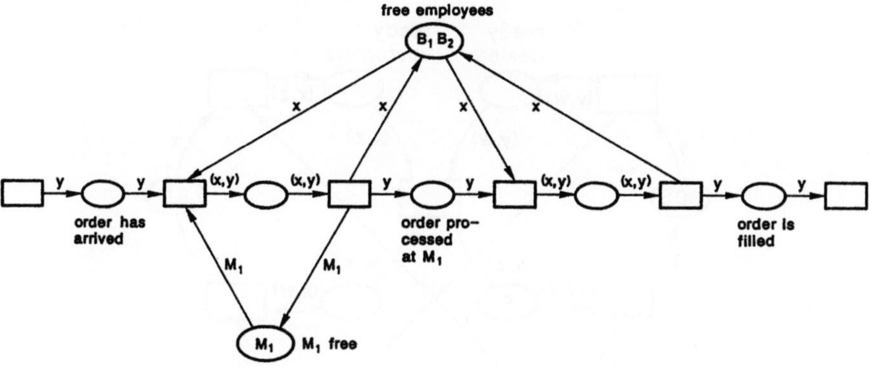

Figure 125

b) We can represent the system of the four seasons with a single loop. To do this we use the 'successor' function which for every season indicates the season that follows, i.e.

$$\begin{aligned}
\mathrm{suc(spring)} &= \mathrm{summer,} \\
\mathrm{suc(summer)} &= \mathrm{fall,} \\
\mathrm{suc(fall)} &= \mathrm{winter,} \\
\mathrm{suc(winter)} &= \mathrm{spring.}
\end{aligned}$$

Figure 126

Problem 24

The special feature of this problem is the fact that we use pairs of things as tokens
that are separated and recombined by the occurrence of a transition.

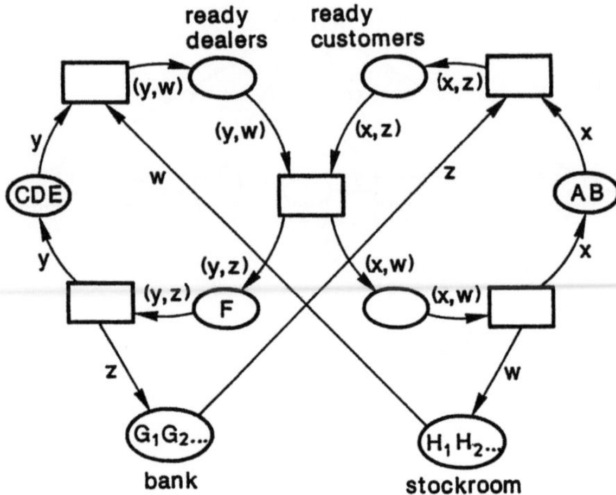

Figure 127

Problem 25

Here variable arrow labels mix with arrow weights for 'black' tokens.

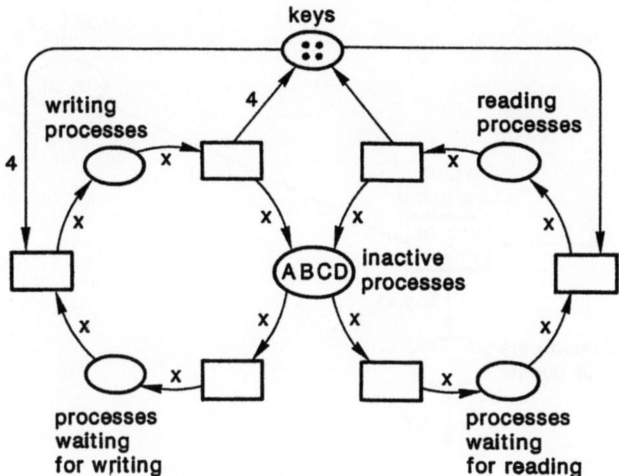

Figure 128

z=B or z=D | cars enter filling station
(x,z)
(A,D) | cars in filling station entranceway
(x,z)
free spaces
drive up to a pump
y=z or y=C
BCD
y
(x,y,z)
unfueled cars at pumps
(x,y,z)
tanks are filled
z
(B,D) | pumps are freed
(x,y,z)
fueled cars at pumps
z
free pumps
(x,y,z)
payment
(w,z)
(w,z)
w
(x,y)
cars ready to leave
w
(x,y)
TU | free cashiers
y
cars leave pumps
x
cars in filling station exitway
x
cars leave filling station

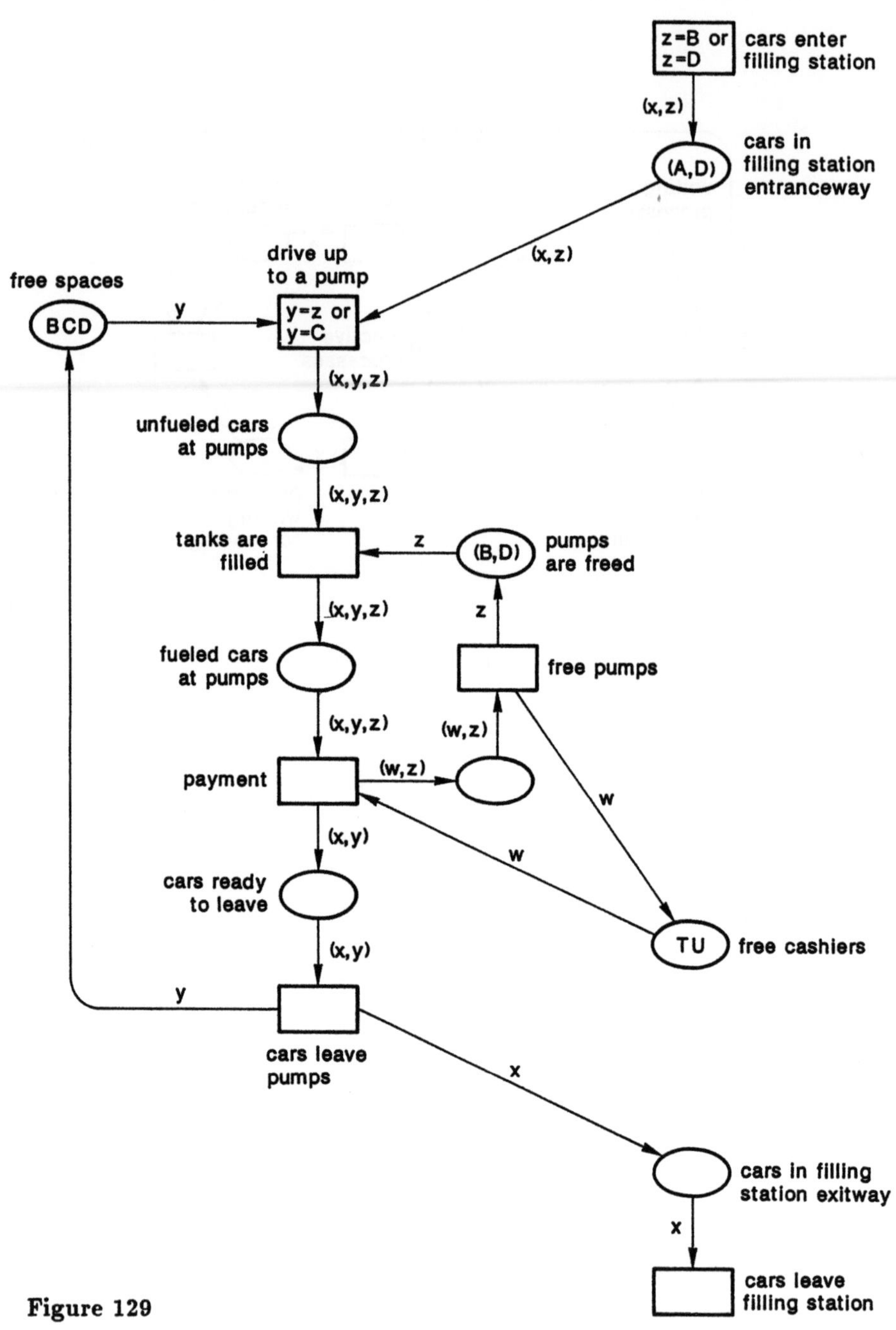

Figure 129

For books we use the variables x and y. For a book x designate an order form on which the book x has been ordered as *order(x)* and the file card for book x as *card(x)*.

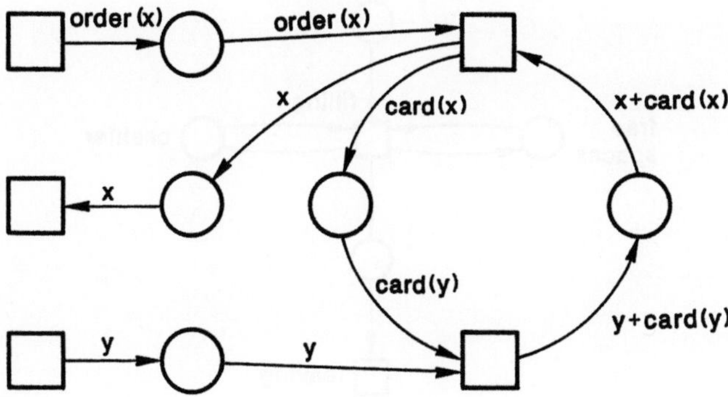

Figure 130

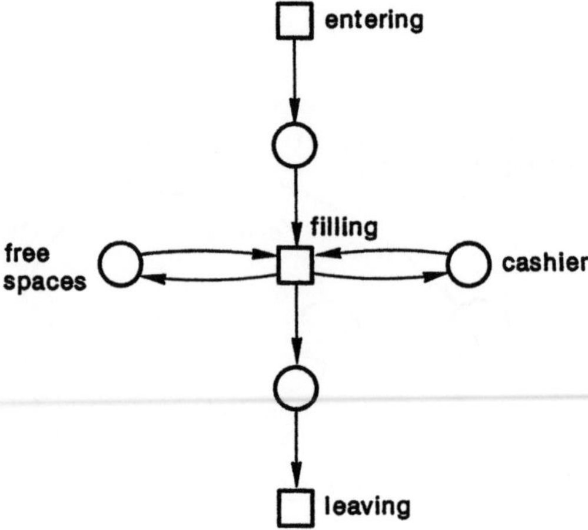

Figure 131

conclusion of
lease without
an agent
landlords
tenants
conclusion
of lease
with an
agent
agent

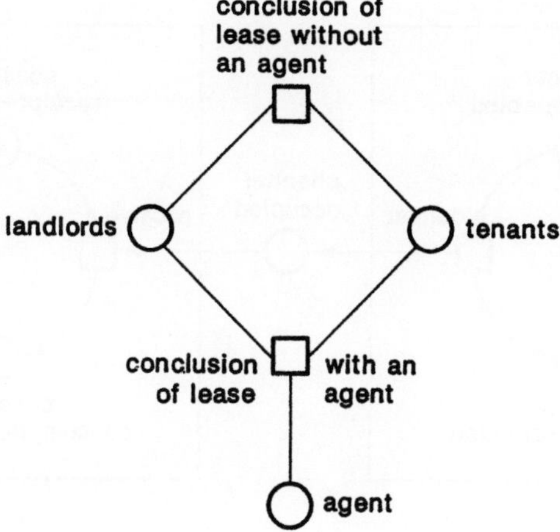

Figure 132

Problem 30

The following refinements can be derived:
a)

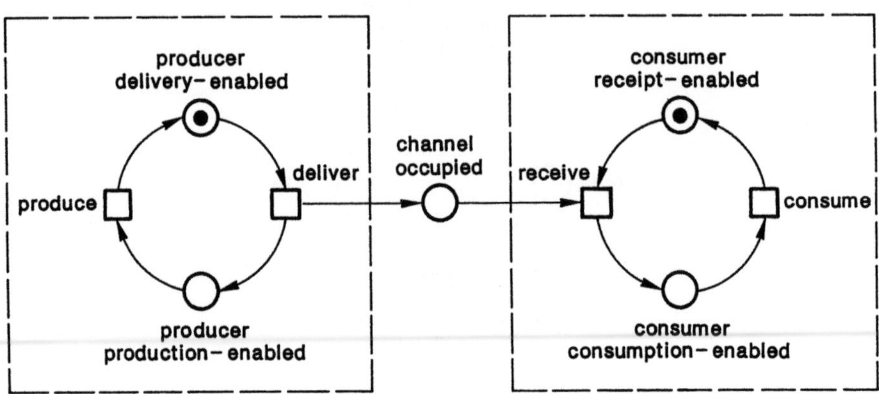

Figure 133

b)

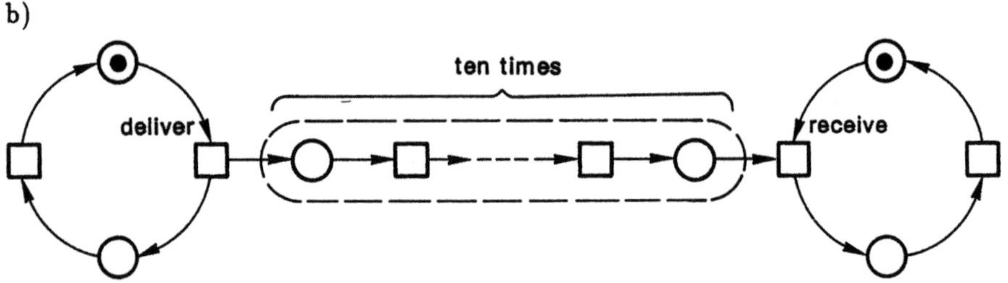

Figure 134

c)

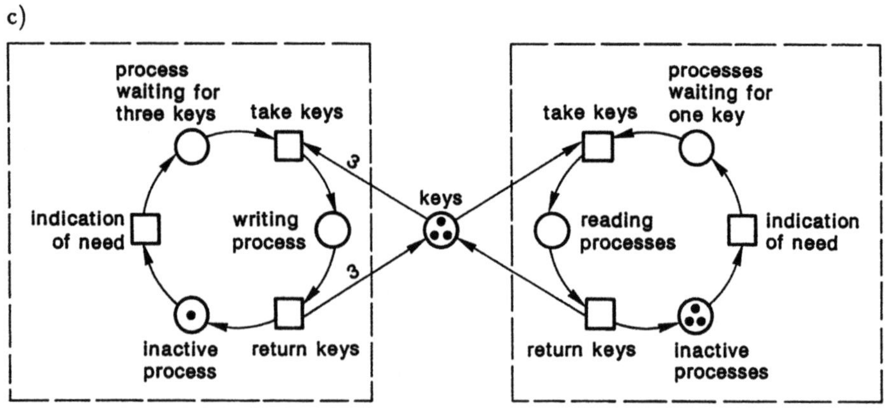

Figure 135

d)

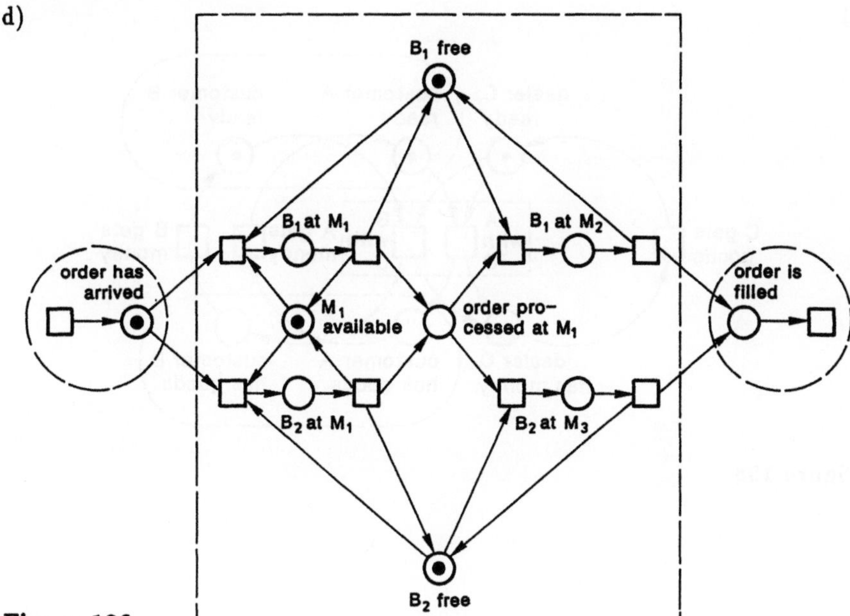

Figure 136

e)

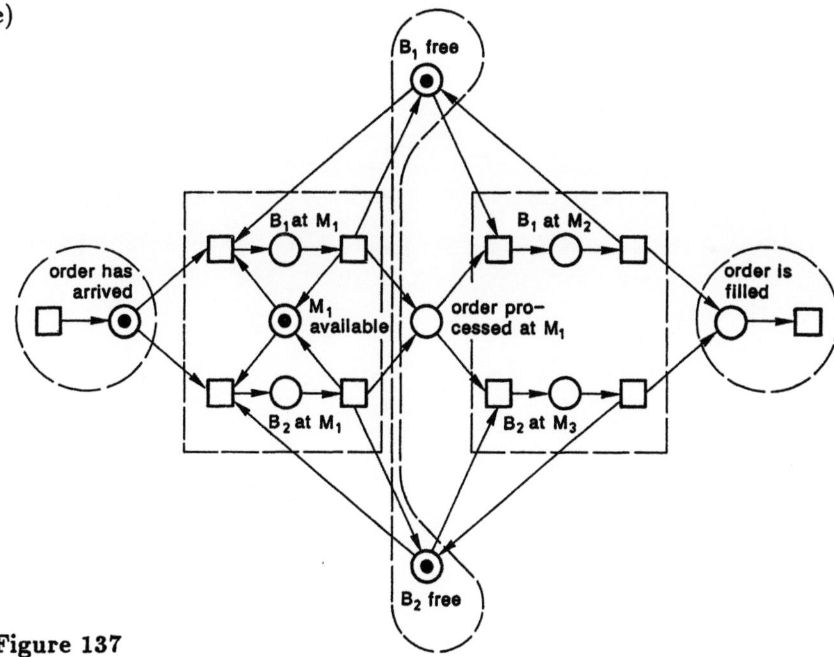

Figure 137

f)

dealer C
ready
customer A
ready
customer B
ready
C gets
goods
A
with
C
B
with
C
A gets
money
B gets
money
dealer C
has money
customer A
has goods
customer B
has goods

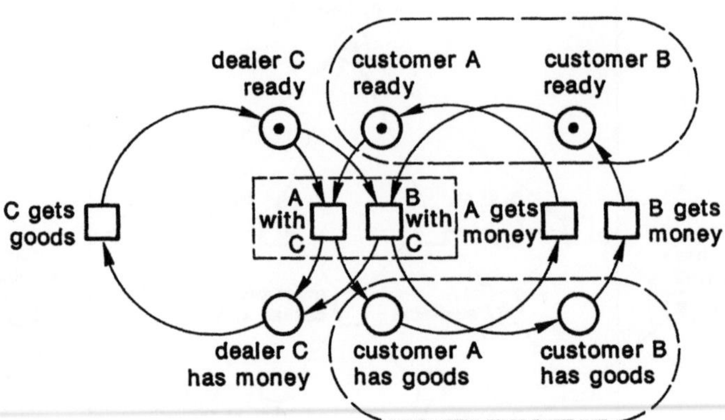

Figure 138

Problem 31

Both refinements are token preserving

Problem 32

a) If we supplement Figure 11 with the condition 'channel occupied' we arrive at Figure 18.

b) If we supplement Figure 18 with the conditions b_0 and b_1 as well as the event e we arrive at Figure 20.

c) If we supplement Figure 27 with a consumer cycle we arrive at Figure 28.

Problem 33

The resulting configuration is not unambiguous, since Figure 97 does not specify whether the goods ordered by letter are to be sent by parcel post or by the company's own courier service. We decide to use the courier service.

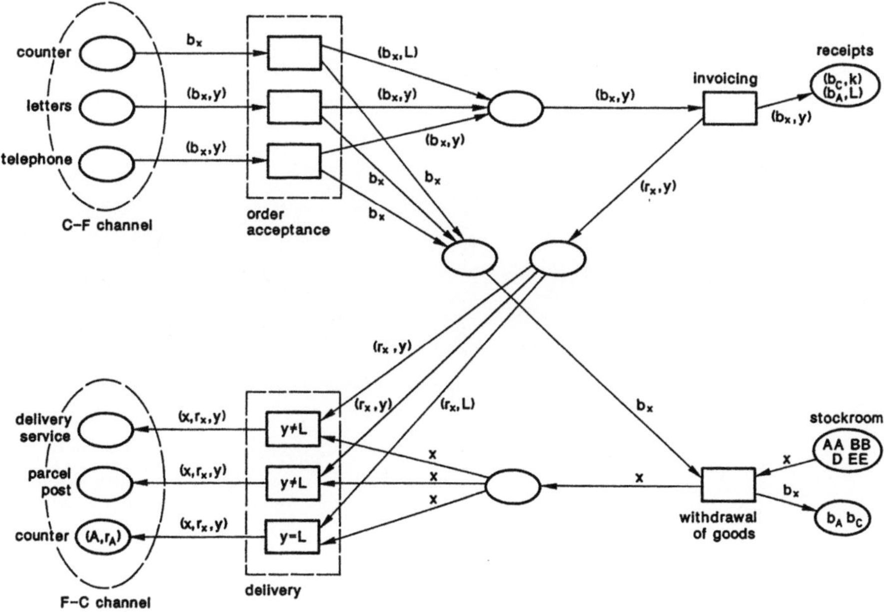

Figure 139

References

a) Textbooks

J. L. Peterson: *Petri Nets and the Modeling of Systems.* Prentice Hall, Englewood Cliffs, New Jersey, 1981

W. Reisig: *Petri Nets – An Introduction.* Springer, Berlin Heidelberg New York Tokyo, 1985

b) Advanced Courses

W. Brauer (ed): *Net Theory and Applications.* Proceedings of the Advanced Course on General Net Theory of Processes and Systems, Hamburg, 1979. Lecture Notes in Computer Science, Vol. 84, Springer, Berlin Heidelberg New York, 1980

W. Brauer, W. Reisig, G. Rozenberg (eds.): *Advances in Petri Nets 1986, Part I: Petri Nets: Central Models and Their Properties, Part II: Petri Nets: Applications and Relationship to Other Models of Concurrency.* Proceedings of an Advanced Course, Bad Honnef, 1986. Lecture Notes in Computer Science, Vols. 254, 255, Springer, Berlin Heidelberg New York, 1987

c) Anthologies

C. Girault, W. Reisig (eds.): Application and Theory of Petri Nets, Informatik-Fachberichte 52, Springer, Berlin Heidelberg New York 1982

A. Pagnoni, G. Rozenberg (eds.): Applications and Theory of Petri Nets, Informatik-Fachberichte 66, Springer, Berlin Heidelberg New York 1983

G. Rozenberg (ed.): *Advances in Petri Nets 1984, 1985, 1987, 1988, 1989, 1990, 1991.* Lecture Notes in Computer Science, Vols. 188, 222, 266, 340, 424, 483, 524. Springer, Berlin Heidelberg New York

d) Bibliography

S. Drees et al: *Bibliography of Petri Nets 1988*, Arbeitspapiere der GMD 315, Gesellschaft für Mathematik und Datenverarbeitung, PO-Box 1240, 5205 Sankt Augustin, Germany

H. Plünnecke, W. Reisig : *Bibliography of Petri Nets 1990*, in G. Rozenberg (Ed.): Advances in Petri Nets 1991. Lecture Notes in Computer Science, Vol. 524. Springer Berlin Heidelberg New York, 1991

e) Periodicals

Petri Net Newsletter, Gesellschaft für Informatik e.V., Godesberger Allee 99, 5300 Bonn 2, West Germany), appears three times a year